RECENT ADVANCES IN MICROBIOLOGY

VOLUME 3

RECENT ADVANCES IN MICROBIOLOGY

Additional books and e-books in this series can be found
on Nova's website under the Series tab.

RECENT ADVANCES IN MICROBIOLOGY

VOLUME 3

N. C. GAUTAM
S. P. TIWARI
AND
RAJESH SHARMA
EDITORS

nova science publishers
New York

Library of Congress Cataloging-in-Publication Data

ISBN: 978-1-53614-057-6
ISSN: 2332-5143

Published by Nova Science Publishers, Inc. † New York

CONTENTS

PREFACE

Microbiology as a science has expanded by leaps and bounds in the past few decades due to advancements in sophisticated instrumentation and recombinant DNA technology, which added a new dimension and revealed an understanding of the subject at the molecular level. Microorganisms are highly ubiquitous in nature and manifold in their activity. Interest in microbiology increased quite significantly due to the wide application of microorganisms in studying different biological processes. An attempt has been made to introduce the readers to several areas of microbiology and microbial biotechnology, and to provide insight into basic concepts of biology and practical applications of microorganisms. To assist the readers in understanding the text, several illustrations and figures have been incorporated in the current volume. The topics covered in this volume are arranged in such a way as to maintain the thread of continuity with Volumes One and Two as well as to make it easy for those who are not acquainted with Volumes One and Two of the series. The editors believe that the previous book of the series, Recent Advances in Microbiology, Volumes One and Two, would also be helpful to readers in strengthening their concepts about different areas of microbiology and microbial biotechnology. The present volume has been complied with the help of leading microbiologists from India and other countries, with an objective of providing current developments in the field of microbiology. This book will be useful for under- and postgraduate students, academicians, researchers and laboratory personnel from disciplines like microbiology, biochemistry, veterinary science, medicine, pharmaceuticals, agriculture, food science, environmental science, industrial microbiology and other allied biological sciences.

Chapter 1 - Microbiology, a branch of biological science, came into existence at a time when several branches of science like Physics, Chemistry, Zoology and Botany were well established. This was due to a lack of facility to observe this form of life. In fact, they were like the ghosts of the scientific world, not visible to the unaided eyes. Although several microbial processes such as curd, vinegar and alcohol production were known from the ancient times, the role of microbes in these processes was not established. These tiny creatures have been playing an important role in every nook and corner of life, ranging from extremely beneficial to severely harmful, from being the source of food and medicines to the ability to cause diseases in plants, animals and human beings. However, here, too, their role was unknown. The society was mainly influenced by the religious-minded and the philosophically-inclined, who believed in the theory of Abiogenesis. According to this theory, life originated from non-living objects. It was widely believed that rotten meat begot flies, frogs came from mud, squirrels were born of trunks of trees, and mice originated from decaying grains.

Civilization and wars were directly affected by microbes; different civilizations were destroyed by a multitude of infectious diseases like typhus, plague, smallpox, syphilis, and cholera, with no weapons to combat them. Wars involve gathering of thousands to millions of soldiers in a place, which leads to unsanitary practices culminating into the spread of infectious diseases. Several soldiers may act as carriers, spreading diseases from endemic areas to new localities. An estimate indicates that the great conqueror Napoleon Bonaparte's army lost more soldiers to disease than to war.

The total credit for establishment of Microbiology goes to two great scientists of their time, Louis Pasteur and Robert Koch.

Chapter 2 - Oxygen is an important gaseous component of the environment and is extremely essential for the living systems. Beyond the physiological levels it may show deleterious effects in the biological system. When oxygen gets converted to partially reduced forms, it results in production of Reactive oxygen intermediates (ROI). Along with this reactive nitrogen species (highly unstable molecules of nitrogen and oxygen), are capable of performing unrestricted oxidation of several cellular macromolecules and producing oxidative damage of cell. These oxidants are either produced exogenously or endogenously. The exogenous sources include exposure to various oxidants in atmosphere like diesel exhaust, cigarette smoke, ozone, nitrogen dioxide, sulfur dioxide, drugs and radiations.

Along with exogenous sources there are several endogenous sources which give rise to various oxidant species inside the biological system. There is production of ROIs during normal metabolism processes. Sources include mitochondria, microsomes and several enzymes like xanthine oxidase, P450 monooxygenase, lipooxygenase, cyclooxygenase, indole amine dioxygenase, monoamine oxygenase etc.

Various inflammatory cells like macrophages, monocytes, eosinophils and neutrophils also generate microbicidal or pathological levels of superoxide ($O_2^{\cdot-}$) through the action of enzyme NADPH oxidase, for microbicidal and other cellular defence mechanism against foreign particles. The superoxide generated is dismutated to hydrogen peroxide (H_2O_2) by the action of superoxide dismutase enzyme. This H_2O_2 is also produced by the action of various other enzymes like monoamine, amino acid oxidase and xanthine oxidase.

Peroxidase enzymes like eosinophil peroxidase (EPO) and myeloperoxidase (MPO) are important in host defense against parasites and pathogens, as they amplify the response of H_2O_2 by forming deleterious HOCl and HOBr. NO is another important free radical produced by the action of enzyme nitric oxide synthase on L-arginine resulting in the formation of L- citrulline and NO. This NO further reacts with superoxide ion to produce more deleterious peroxynitrite radical which causes damage to tissues. In plants exposure to drought stress, salt stress, chilling, heat shock, heavy metals, mechanical stress, nutrient deprivation, pathogen attack and high light stress also result in ROI generation via photorespiration, mitochondrial respiration, glycolate oxidase pathway in peroxisomes and NADPH oxidase system.

To combat with these reactive species body contains a strong defense mechanism consisting of several antioxidants. Antioxidants are defined as the substance, that even present in low concentrations (compared to those of an oxidisable substrate), significantly delays or prevents oxidation of that substance. The concept of antioxidants *in vivo*, includes antioxidant enzymes, iron binding and transport proteins, other compounds affecting signal transduction and gene expression. These are also of two types:

a) Non enzymatic antioxidants - These consists of urate, reduced GSH, vitamin E and C, extracellular SOD, catalase, GSH-Px, GSH, mucopolypeptide glycoproteins, cerruloplasmin, Fe binding proteins e.g., lactoferrin and transferrin and bilirubin.
b) Enzymatic antioxidants - Along with the non enzymatic antioxidants several enzymatic antioxidants are also required which either act alone or synergistically against oxidative stress. The major enzymatic antioxidants present in lungs are superoxide dismutases, catalase, glutathione peroxidase and small molecular weight redox proteins (thioredoxins, peroxiredoxins and glutaredoxins, heme oxygenase-1) which play a crucial role in the lung antioxidant defence mechanisms.

The antioxidants present in cells, cell membranes and extracellular fluids are up-regulated and mobilized to neutralize the excessive ROS formed during oxidative insults. Any deficiency in the antioxidant defense of the body, to combat with these oxidants produced, leads to oxidative damage to various macromolecules (protein, lipids, DNA) and this condition of imbalance is known as oxidative stress. Oxidative stress has been found to be involved in the pathophysiology of various diseases including neurodegenerative diseases like Alzheimer's disease, Parkinson's disease, diabetes, rheumatoid arthritis, cardiovascular diseases, lung inflammatory disorders like asthma acute respiratory distress syndrome, chronic obstructive pulmonary disease and neurodegeneration in motor neuron diseases. Antioxidants acts to reduce/inhibit oxidants by various mechanisms like:

a) They inhibit the formation of free radicals in the initiation step of auto-oxidation
b) Interrupt the propagation of the free-radical chain.
c) They can chelate free metal ions in solution
d) Act as reducing agents (electron donors) and remove or reduce concentrations of oxygen.

For many decades, there has been a growing interest in finding novel antioxidants. Natural antioxidants isolated from plants, fungi and marine algae represent the most useful nutraceuticals and functional foods for health protection and disease prevention. In the search of new natural antioxidants, a number of bioactive compounds obtained from plants, algae and microorganisms have been demonstrated to possess potent antioxidant activity. Further, there is an overview of some microbes or their secondary metabolites which are studied for their antioxidant potential.

Chapter 3 - Vanilla is the world's most popular flavoring agent for sweetened foods. Pure vanilla extracts are widely used for flavoring in ice creams, chocolates, tobacco, baked foods, puddings, cakes, liquors and as a fragrance ingredient in perfumery. It improves the flavor of caramel, coffee and some dairy products. It is also used in spice oleoresin formulations, sausages, seasonings etc. Natural vanilla is a complex mixture of flavor components extracted from the cured beans of the *Vanilla* plant. More than 170 volatile aromatic components have been identified, apart from the principal components vanillin, vanillic acid, *p*-hydroxybenzaldehyde, vanillyl alcohol, *p*-hydroxybenzoic acid and *p*-hydroxybenzylalcohol that are present in detectable amounts in the *Vanilla* pod. Natural vanillin also acts as an antimicrobial agent by inhibiting molds and yeast in fruit purees and fruit-based agar systems.

Chapter 4 - The Plant Growth Promoting Rhizobacteria (PGPR) are bacterial species that live in the rhizosphere (root zone) of a plant and increase plant growth. This group includes all the bacteria that help plant growth either directly or indirectly, through various means. The term PGPR was used to describe the bacteria that colonize plant roots following inoculation on seeds to enhance plant growth. Plants interact with an enormous volume of microorganisms by way of their root system, due to the richness of organic matter in the soil. Having existed in the proximity of roots since times immemorial, microorganisms have evolved various strategies to interact with plants for availing nutrients. Some microorganisms interact by colonizing the root surface to avail root exudates. Others interact strongly by secreting a spectrum of metabolites and can modify the rhizosphere properties. These changes, in turn, affect plant growth directly or indirectly. Possible ways by which the microbiota can interact with plants are by (i) providing nutrients to plants, (ii) growth stimulation of plants by secretion of growth hormones, (iii) controlling plant pathogens, (iv)improving soil quality, (v) minimizing pollutants through bioremediation and by (v) bioaccumulation or microbial leaching of inorganic compounds. It is generally considered that rhizosphere is dominated by non-fermenting, Gram-negativerods, especially pseudomonads.

The phenomenon of plant growth promotion is governed by a variety of bacterial traits and specific genes, which include chemotaxis toward seed and root exudates, motility, production of pili or fimbriae, production of specific cell surface components, ability to use specific components of root exudates, protein secretion and quorum sensing. Latest screening methods based on gene fusion technologies are aiding the identification of new, previously uncharacterized genes. Altering the mutants for expression of traits is helping in understanding the exact role of each gene. These methods use reporter transposons and *in vitro* expression technology (IVET) to detect genes expressed during colonization.

Inoculant survival in the rhizosphere is a prerequisite for successful plant growth promotion. Often the soil inoculants that perform well in laboratory fail to produce satisfactory results in the field. These challenges restrict successful commercialization of PGPR as most of the available agro-industrial rhizobacterial strains face population decline after their introduction in soil. Also, various soil types, ranging from fertile to non-fertile ones exist across the globe; therefore, the strain successful in one place may not perform equally well in another patch of land. The approach of selectively screening advantageous native soil microflora and its mass scale reintroduction in the rhizosphere can aid in enhancing bioinoculant survival period.

A wide range of bacteria was successfully tested to increase plant growth. Some bacterial genera like *Alcaligenes*, *Arthrobacter*, *Azospirillum*, *Azotobacter*, *Bacillus, Burkholderia*, *Enterobacter*, *Klebsiella*, *Pseudomonas* and *Serratia* are well known to enhance plant growth. Their commercial formulations as bioinoculants are available for use on agricultural lands. However, the majority of literature has focused on *Pseudomonas* and *Bacillus*. The *Pseudomonas* genus has evolved as a promising rhizobacteria, which enhances plant growth by several mechanisms, enabling them to utilize various substrates released by the roots. Microorganisms like *Pseudomonas putida* are competitively more stable in rhizosphere because they secrete heterologous siderophores. Often such inoculants perform well under nutrient deficient conditions. Carbon-starved *Pseudomonas fluorescens* R2f populations have been reported to enhance resistance to otherwise lethal treatments, like exposure to ethanol, high temperature, osmotic tension and oxidative stress. Most of such non-sporing PGPR require special carriers for survival at the time for their mass cultivation. However, working with spore-forming populations is more convenient due to their higher survival rate and easier mass preparation. The genus *Bacillus*, an endospore producer, is tolerant to desiccation and heat. It has a longer life in commercial formulations and is a genus of choice in the biofertilizer industry. Bacterial spore-formers have short generation time, greater mobility and the capability to colonize roots. Moreover, they produce a wide variety of secondary metabolites including plant growth-regulating substances and antibiotic compounds.

Chapter 5 - Milk is defined as the white lacteal secretion obtained by complete milking from the herd of healthy milch animals, five days after calving and 15 days before calving. Milk is valued because it is an important source of many of the nutrients essential for the proper development and maintenance of the human body, and one of the most nutritious foods. It is rich in high quality protein providing all essential amino acids. It contributes to total daily energy intake as well as essential fatty acids, immunoglobulins, and other micronutrients. It is the primary source of nutrition for young mammals before they are able to digest other types of food. First lactation is known as colostrum, which carries the mother's antibodies to the baby, improving the immunity of neonates and preventing the risk of many diseases. Cow's milk is the predominant type of milk consumed in several countries, although goat, buffalo, sheep, and camel milk are also consumed.

Wide ranges of milk products are available throughout the world, including curd, cheese, milk powder, ice cream, butter, fermented products etc. The precise nutrient composition of raw milk varies from species to species, and by a number of other factors, but it contains significant amounts of saturated fat, protein and calcium as well as vitamin C. A recent study states that approximately 73% of the calcium, 33% of the phosphorus, 31% of the riboflavin, 21% of the vitamin B12, 19% of the protein, 16% of the magnesium, 17% of the vitamin A, 10% of the vitamin B6, and 6% of the thiamin in addition to appreciable amounts of vitamin D and niacin is contributed by dairy-related food,. Thus, milk and dairy products act as nutrient-dense foods providing abundant amounts of protein, vitamins and minerals necessary for growth and development. A number of proteins found in milk under various conditions exhibit antimicrobial activity. For example, immunoglobulins (antibodies) are protective proteins that are important in the transfer of passive immunity from the mother to the neonate. The immunoglobulins protect the neonate from infection until their own immune system is developed. Immunoglobulins are a component of the natural defense mechanism. They are synthesized in response to the presence of foreign particles, such as bacteria and viruses. The synthesis process is specific for the foreign particle that is present, resulting in an immunoglobulin structure that is capable of recognizing the foreign particle and eliminating it from the body. Immunoglobulins are found in high concentrations in colostrum, the first milk, and in low concentrations in milk. In addition to its value as a nutrient source, the purpose of this chapter is to critically review the antimicrobial properties of milk and milk products.

Chapter 6 - Microbial life has evolved the capacity to thrive in a broad range of thermal environments. Thermal adaptation, in particular to extreme conditions, has restricted the range of tolerable temperatures. It seems to be a fundamental law that, wherever microbial life can survive, it will be found to exist as they are the most hardiest, spore forming organisms which can grow and reproduce at temperatures less than 0°C in high-altitude cloud droplets. The largest proportion of biomass on earth is generated at cold temperatures (≤5°C). These environments comprise, among others, the Antarctic, Arctic and mountain regions, as well as the deep-sea waters that cover three-quarters of the planet surface. This is mainly due to the contribution of vast numbers of microorganisms in the oceans, although the cold biosphere extends to permanently cold alpine regions, caves, upper atmosphere, and polar regions, in addition to seasonally cold environments.

Evolution has allowed these cold adapted organisms, called psychrophiles, not to merely survive, but to breed and grow successfully in the restrictive conditions of cold habitats.

Far less is known about psychrophilic bacteria than about thermophiles, and it is not even certain yet whether life on earth originated in a hot or cold environment.

Psychrophiles display metabolic fluxes at low temperatures that are more or less comparable to those exhibited by closely related mesophiles living at moderate temperatures, clearly showing that mechanisms of temperature adaptations are involved.

Cold-adapted microorganisms can be categorized as psychrophilic (an organism with an optimal growth temperature at or below 15°C, and a maximum growth temperature below 20°C), psychrotrophic (an organism exhibiting the ability to grow at temperatures below 15°C but exhibiting maximum growth rates at temperature optima above 18°C) organisms, and are often subjected to other extreme environmental parameters and psychrotolerant (organisms tolerant of low growth temperatures). Some are specialized to very low temperatures (steno-psychrophilic bacteria, with optimal growth temperature <15°C), and some tolerate temperature up to 20°C (eury-psychrophiles). The differentiation between psychrotrophs and psychrophiles is, by and large, ignored now-a-days and all the cold-tolerant bacteria are called psychrophiles.

These microorganisms are devoid of any temperature regulation, their internal temperature synchronizes to that of the surrounding environment. They have very strongly challenged the strong negative effect of low temperatures on biochemical reactions, and continue to breed, grow and move at rates similar to those achieved by closely related species living in temperate environments.

Lot of important and essential adaptations are required to survive at such low temperatures, including the expression of enzymatic activities at appropriate levels as well as the maintenance of membrane stability and permeability, the latter being essential for many of the major cellular functions such as nutrient uptake, passive and active transport of various compounds, electron and proton transport, photosynthesis, environmental sensing, and recognition processes. The changes at the level of three dimensional structure of protein are incorporated as mutations or modifications at the gene level. These differences via transcription and translation form an altogether differently modified pattern of proteins expressed through the thermal regulation of gene expression.

Chapter 7 - Single cell protein (SCP) refers to dried microbial cells or total protein extracted from pure microbial biomass (algae, bacteria, filamentous fungi and yeasts), which can be used as human food (food grade) or animal feed additive or supplement. Due to rapid growth in human population, deficiency of nutrients, especially protein, has been observed in human as well as animal diet. Single cell protein have multiple applications in animal nutrition, such as fattening of calves, poultry, pigs and fish breading. In food, SCP is used as aroma carrier, vitamin carrier and emulsifying aid, apart from improving the nutritive value of baked products in ready-to-serve meals in diet recipes. Also, SCP has found usage in the manufacturing sector, in paper processing, leather processing and as foam stabilizers.

The term single cell protein (SCP) was coined in 1966 by Carol L. Wilson at the Massachusetts Institute of Technology (MIT), U.S.A. Rising populations and the need for relieving protein malnutrition in developing countries has catalyzed several efforts to intensify the production of protein both from conventional agriculture as well as non-conventional sources. Microbial biomass/cells or cell cultures like algae (*Spirulina*), yeasts/fungi (*Saccharomyces cerevisiae*) and bacteria (*Methylophilus methylotrophus*) as a source of SCP has attracted the attention of biologists and technologists as promising options to reduce the protein demand-supply gap. The greatest promise of SCP, particularly in India, lies in its potential as animal and poultry feed and also as protein supplement for our largely vegetarian consumers.

It has been estimated that if necessary measures are not taken, the malnutrition problem will lead to some major health crises in the developing countries. Therefore, it is very important to increase protein production and also make it available to the general population by utilizing all the available methods. The increasing global demands for food and feed protein spurred the search for non-conventional protein sources. A great deal of research interest has been focused on the potential of converting agricultural wastes to microbial protein.

Chapter 8 - Food and Agriculture is the largest sector in the national economy. Before reaching the consumer, food passes through many hands and is handled and processed in several ways. To retain its quality, it is essential to ensure the integrity and safety of food throughout the supply chain. Measures taken to ensure the above will also add to the value of food as a commodity and lead to better acceptability from the consumer, both domestic and international. According to the reports released by the World Health Organization (WHO) and Food Agriculture Organization (FAO) of the United Nations, illness due to contaminated food is one of the most widespread health problems in the world and an important cause of reduced economic productivity, especially in developing and under-developed countries. Currently, millions of people worldwide are suffering from diseases caused by contaminated food, which inflicts a heavy social and economic cost. Major challenge of spoilage in the food industry is due to microbial contamination. While there is no easy solution to this problem, developing rapid tests for detection of food borne pathogens can be a significant contribution. Growth in consumer demand for safe food has increased interest in the establishment of an accurate and rapid methodology for the detection of contamination in food products, which should be cost- effective, simple, specific and capable of automation for online monitoring.

Chapter 9 - Several microorganisms are capable of infecting the brain and its associated tissues. Bacteria and viruses are the most common microorganisms that infect brain. Fungi, parasites and others can also infect the brain, though rarely. Germs infect the brain, causing an inflammation of the invaded area and manifesting the disease. The nomenclature of the brain infections vary, depending upon the exact part of the brain that has been targeted by the microorganism:

Meningitis- Meningitis is the inflammation of the meninges, which is the three-layered membrane enclosing the brain and spinal cord; the fluid it is bathed in is called the cerebrospinal fluid (CSF).

- *Encephalitis*- It is an inflammation of the brain itself.
- *Myelitis*- Spinal cord inflammation is known as Myelitis.
- *Abscess*- Accumulation of infectious material and offending microorganisms within the CNS.

Meningitis may be either of viral or bacterial origin

1. *Viral meningitis* requires no specific treatment, and goes away completely without complications.
2. *Bacterial meningitis* is a very serious disease and may result in speech faults, hearing loss, seizures, loss of extremity function or deduction, permanent brain damage, and even death. Bacterial meningitis is of great concern worldwide and may be a serious threat to global health.

Chapter 10 - One of the major health care problems of human beings is infectious diseases. The majority of these diseases are caused by contaminated food, water and air. They are responsible for significant human pathogenesis, morbidity and mortality around the world. Infectious diseases initiated in a particular region of the world can rapidly spread around the globe. The most effective strategy to prevent the spread of an infectious disease is rapid and accurate identification of the pathogen. Early diagnosis helps in control and eradication of the infectious disease. It also gives sufficient time to government and clinicians to take necessary preventive measures for disease-zoning. On the other hand, misdiagnosis and delay in diagnosis of potentially dangerous pathogens is a limitation in early control of the disease. During the last century, significant advancement has been made in disease diagnostic techniques, with the most common methods used for identification or detection of infectious agents being microbial culture and colony counting, immunological assays (immunoassays) and molecular methods.

Chapter 11 - Soil contains a plethora of microorganisms, with both beneficial and deleterious attribute constituting an environmental surrounding termed as "soil micro-flora". Soil containing beneficial microbes is a boon for agricultural crops, but existence of harmful microorganisms can be detrimental to both plant and animal health. Diverse population of microorganisms like fungi, bacteria, protozoa, algae and actinomycetes constitute a specific soil micro-flora of a specific region.

Vegetables are the major component of nutritional and livelihood security. As the production of crops depends largely on soil micro-flora, soil rich with beneficial micro-flora/microorganism is known to enhance crop productivity. The physico-chemical nature of soil is also dependent upon the presence of microorganisms in the soil. The population of microorganism also reflects the fertility status of specific soil. Productivity of vegetables per unit area is diminishing and there is a necessity to focus on soil environment. Pathogenic soil microorganisms play a major role in decreasing vegetable production and effective management of them is a challenge. The present chapter deals with several soil microorganisms viz. *Fusarium* sp., *Verticillium* sp., *Pythium* sp., *Rhizoctonia* sp., *Sclerotium* sp., *Sclerotinia* sp. and *Ralstonia* sp. which are harmful to soil health causing several diseases like wilt, damping off, rot etc. on a wide variety of vegetable crops. An effort is also made to construct a roadmap for their effective control.

In: Recent Advances in Microbiology. Volume 3 ISBN: 978-1-53614-057-6
Editors: N.C. Gautam, S.P. Tiwari and R. Sharma

Chapter 1

From Leeuwenhoek to Craig Venter

***R. Sharma*[1], *Rishi Srivastava*[2], *R. K. Singh*[3] *and S. P. Tiwari*[2*]**
[1]Department of Biotechnology, VBS Purvanchal University, Jaunpur, India
[2]Department of Microbiology, VBS Purvanchal University, Jaunpur, India
[3]Department of Microbiology, HNB Garhwal University, Srinagar, India

Introduction

Microbiology, a branch of biological science, came into existence at a time when several branches of science like Physics, Chemistry, Zoology and Botany were well established. This was due to a lack of facility to observe this form of life. In fact, they were like the ghosts of the scientific world, not visible to the unaided eyes. Although several microbial processes such as curd, vinegar and alcohol production were known from the ancient times, the role of microbes in these processes was not established. These tiny creatures have been playing an important role in every nook and corner of life, ranging from extremely beneficial to severely harmful, from being the source of food and medicines to the ability to cause diseases in plants, animals and human beings. However, here, too, their role was unknown. The society was mainly influenced by the religious-minded and the philosophically-inclined, who believed in the theory of Abiogenesis. According to this theory, life originated from non-living objects. It was widely believed that rotten meat begot flies, frogs came from mud, squirrels were born of trunks of trees, and mice originated from decaying grains.

Civilization and wars were directly affected by microbes; different civilizations were destroyed by a multitude of infectious diseases like typhus, plague, smallpox, syphilis,

[*]Corresponding Author Email: sptiwarimicro@gmail.com.

and cholera, with no weapons to combat them. Wars involve gathering of thousands to millions of soldiers in a place, which leads to unsanitary practices culminating into the spread of infectious diseases. Several soldiers may act as carriers, spreading diseases from endemic areas to new localities. An estimate indicates that the great conqueror Napoleon Bonaparte's army lost more soldiers to disease than to war.

The total credit for establishment of microbiology goes to two great scientists of their time, Louis Pasteur and Robert Koch.

Keywords: Leeuwenhoek, Craig Venter, Marcello Malpighi, Lazzaro Spallanzani, Franz Schulze, Pasteur, Koch, Tyndall, Magic Bullet, Genner, Petri, immunization

Discovery of Microbes

During the 17th century, a different journey was started in the field of Biology, when it was realized that the living components are made of individual living units known as cells – this was after the discovery of cells by Robert Hooke. The foundation stone for the evolution of microbiology started with the discovery of microscope for the first time by Zacharias Janssen, and his father, Hans, a Dutch spectacle-maker between 1590 and 1610. Marcello Malpighi (1628–1694), an Italian scientist, used the microscope for the first time for studying biological materials. Another microscope of that time was designed by Antony van Leeuwenhoek (1632–1723), which was of a high resolution and could be used to observe different forms of microbial life. Antony van Leeuwenhoek was born in Holland in 1632. Interestingly; he did not attend any university and worked as an officer to calibrate wine. Using his microscope, he was the first to observe the microbial form of life and named them as animalcules. He was of the opinion that this form of life could originate from existing organisms of the same kind and hence, can be considered the first opponent of the theory of abiogenesis. In his letters sent to the Royal Society, which were published in Philosophical Transactions, demonstrated the existence of molds having club-shaped structure arising from the vegetative body now termed as sporangiophore and sporangia. After his first report, he never looked back and sent nearly 200 letters to the society; in 1680, he was elected as a member of the Royal Society. He observed and reported different shapes of microorganisms, such as circular (coccus), rod (*Bacillus*) and spiral forms as well as protozoa, yeast, bacteria, and algae, and compared their size with sand and red blood corpuscles; he also studied their motile behavior. Although the fabrication of microscope constructed by him was simple, its operation was an intriguing mystery to the scientific community. Due to his immense contribution in the discovery of these tiny creatures, he must be remembered as the Father of Microbiology. However, despite the report of presence of microbes in the environment, their existence was not accepted because of the dominance of the followers of Abiogenesis (Dobell, 1932; Stanier, 1987; Staley et al. 2007).

Conflict over the Theory of Abiogenesis

As described earlier, there was considerable conflict regarding the evolution of life. Major groups were of the opinion that life had originated from non-living objects, a theory known as Abiogenesis. Francesco Redi (1626–1697) was an opponent of this opinion and he conducted an experiment using meat, a container and muslin cloth; one set was left open while the other covered with muslin cloth. In the open container, flies developed while in the covered container, appearance of flies was not noticed. He was of the view that flies can develop only from eggs. However, the role of microbes in fermentation and putrefaction was not established, and hence, his results were not accepted.

Lazzaro Spallanzani (1729–1799) conducted a series of experiments to confirm the report of Redi. He used heat to sterilize normal pond water in sealed glass containers. This process inhibited the appearance of life in the infusions. However, John Needham, an English cleric and proponent of spontaneous generation opposed his idea by saying that a vital force essential for growth of life was destroyed by heat and sealing of containers. In contrast, Spallanzani was of the opinion that heat destroyed all the vegetative cells present in the infusion but was not capable of killing the spore-formers for a limited period of heating. Antoine Lavoisier, who discovered oxygen and reported that oxygen is necessary for life, opined that oxygen was destroyed by heat and thus, life did not appear in the infusion.

Franz Schulze (1815–1873), a German chemist in 1836 criticized the report of Antoine Lavoisier, and opined that oxygen was not the key factor. He designed a glass jar where a glass tube was inserted in the mouth of a stoppered flask. This glass tube was joined with a bulb containing acid and another containing potassium hydrate. The flask containing infusion was heated for sterilization. He drew air continuously through the bulbs and observed that the infusion remained sterile. A control flask containing direct entry of air was found to express growth of microbes. He suggested that flasks containing infusions could be sterilized and contaminated through the entry of air.

He suggested that infusion could be sterilized; however, till then, the process of fermentation had not been reported. The credit for discovering fermentation goes to three scientists. Working in 1836, Charles Cagniard-Latour (1777–1859) correlated the role of yeast in alcohol and beer fermentation, which is produced by their biological activities. Theodor Schwann (1810–1882), a German physiologist discovered the yeast cell in 1837 and suggested it to be a plant kingdom member. He termed yeast as "zuckerpilz" (sugar fungus). Further, Friedrich Kützing (1807–1893) suggested that this process was caused by living organisms and different types of products are produced by different organisms. Louis Pasteur started working on fermentation and suggested that organic living beings are present in the ferment and these are capable of reproducing and causing chemical changes. He christened them 'germs,' which are present in the air and are responsible for fermentation. These germs were fungi and bacteria. He used the aspirator filter system to recover microbes. He utilized the knowledge of H. G. F. Schröder and T. von Dusch to use spun cotton-wool, which could efficiently filter the microbes. He drew air through this cotton and dissolved it in a mixture of alcohol and ether; he observed it under a microscope and confirmed the presence of numerous round and oval bodies. Quantitative variation was observed in the number of these germs, with respect to temperature, moisture and direction of air (Strick, 1999).

He then designed several flasks with specialized necks of various shapes (up and down), added nutrients in distilled water for the growth of these germs, and boiled this infusion for sterilization. A platinum tube was attached to the neck, which was heated till red hot, and air was withdrawn through it into the flask. He noticed that air that was withdrawn through this tube remained sterile and no growth was observed in the flask. The flask was similar in shape to a swan's neck, having bends where particles were trapped and the air flowing in was rendered free of germs. He observed that when the neck of the flask was broken and exposed to air, the infusion became turbid with the growth of germs. Swan-neck flask had unencumbered passage of air and when air passed through, it settles down at position where tube was bended. Based on this, he concluded that the germs existed in air and when they came in direct contact with the nutrient-rich infusion, they started growing in it. He proved that life could only emerge from pre-existing life. He repeated the experiment by exposing the broth at different locations. He found qualitative and quantitative differences in microbial population at different sites. In 1861, his work appeared for the first time as a treatise on the organized bodies existing in the atmosphere (Dobell, 1932; Stanier, 1987; Ketchum, 1988; Krasner, 1995; Staley et al. 2007).

PROBLEMS WITH BACTERIAL SPORES

Pasteur was of the opinion that boiling of media for 1 hour could kill the microbes, but when hay was boiled for 1 hour, it did not abolish the possibility of appearance of life. The experiment succeeded with yeast extract broth (Krasner, 1995). The scientific community started hunting for the solution of this question. Ferdinand Cohn (1877) found that certain structures were produced in the cells; he called them endospores and found them to be resistant to heat. These could easily be observed under the microscope as refractile bodies, which contaminated alkaline hay infusions and produce normal rods; thus, endospores could produce vegetative cells. The science of microbiology was, therefore, advancing; however, how to kill the spores still remained an unsolved riddle. The idea to overcome this problem was given by the English physicist John Tyndall (Tyndall 1877; Drews, 2000). He conducted several experiments to observe the effect of heat on the destruction of these spores. In one of his experiments, he took fresh vegetable and meat infusion in tube and boiled it for 5 min. In another set, he took the infusion from dried hay and boiled it for 5 min. He compared the results and found that in the second set, heating was ineffective because it contained spores that could not be killed by a mere 5 min. boiling; in fact, even when the time of boiling was increased to one hour, microbes appeared and spores were not destroyed. He was of the opinion that bacteria existed in two phases: one was thermolabile while the other was thermotolerant. Thus, this confirmed the idea of the German botanist F. Cohn, who had talked of resting bodies such as spores. Tyndall further developed a technique for this problem wherein the infusion was sterilized by discontinuous heating. He suggested that alternate boiling and cooling allowed the spores to germinate, killing them in the second boiling step. He repeated the cycles of boiling and cooling for 5 times to make the infusion sterile. This process was termed as Tyndallization. Hence, this brilliant work of Pasteur and Tyndall proved that life can come from pre-existing life, and disproved the theory of abiogenesis.

France being the largest producer of wine, was struggling with the spoilage of wine. King Napoleon asked Pasteur to find a solution for this problem. In response, he developed the technique of Pasteurization. He suggested that fermentable solutions can be prevented from spoilage by this process, which involves a special heating treatment. This method was particularly implemented to save wines and beers from spoilage, by heating at 55°C.

Clinical Microbiology and Culture Techniques

First report on the cause of diseases and that "contagions" were responsible for their spread was made by Girolamo Fracastoro (c.1478–1553).He wrote that the diseases spread through direct contact, air and clothes.

Culture Techniques

The exact agent responsible for the disease was not known and there were disputes regarding the cause of disease when the concept of abiogenesis was popular. The involvement of a living organism in disease was not established. For the first time, Louis Pasteur reported the Germ Theory of Diseases while he was working on fermentation (Krasner, 1995).He proved that germs existed in the air and were responsible for contamination. These germs survived in the body and reproduced there, before transmitting to another host. During this period, the exact condition for growth and multiplication was not established, which was only possible if pure culture techniques were established. Brefeld (1872) achieved success in growing fungal mycelia from spores on gelatin, while the German botanist Schroeter found growth of pigmented bacterial colonies on slices of potato in humid environment. This could be considered the first efforts of solid media-based pure culture techniques.

The credit for the development of this technique goes to Robert Koch (1881), who for the first time developed a technique for the isolation and culture of bacteria. During his experiment, he took a sterilized container closed with mold-proof cotton, and sterilized nutrient was poured into this container. Now, this nutrient solution was inoculated with microbes to yield pure culture. Full growth was observed in the container after sufficient incubation. After this, a small volume of culture was inoculated into fresh sterilized container, which exhibited luxuriant growth post incubation. He was of the opinion that frequent repetition of transfers to fresh medium resulted in obtaining pure culture. However, there were several doubts regarding his experiments:

1) Whether the container used for growth was totally sterilized
2) Whether the cotton used was resistant to mold
3) Whether the inoculums contained any unwanted microbe
4) Whether the nutrient solution used for growth was totally sterilized
5) Whether during transfer, any foreign contaminant entered

For the study of pathogenic microbes, pure culture technique was essential. Dr. Robert Koch's experiments showed the light here. He took a boiled piece of potato and its surface was exposed to a moistened chamber to prevent its drying up. Within two to three days, small pin-head colonies like droplets of various colors such as yellow, brown, light grey and red appeared on the exposed surface, which gradually increased in size, finally leading to the disintegration of the potato slice. Another set of experiments was performed wherein a hot knife was used to remove the cover of potato and soil adhered to it, the slice was placed in a moist chamber with restricted air. Appearance of colonies was not observed and the potato slice remained unchanged. This indicated that colonies could only develop when the potato slice came in contact with air. Experiment revealed that dust particles contained more than one microbe and when broth was used for culture, individual colony could not be formed, as observed on potato slices. Hence, solid nutrient media was essential for pure culture. When the colonies growing on boiled potato slice were transferred to another slice, the appearance of similar colonies was observed but the bacteria pathogenic to animals did not appear on potato slice.

Louis Pasteur used different infusions to observe the growth of microbes, but it was noticed that a single suitable infusion was not suitable for all types of microbes (Krasner, 1995).Dr. Koch and his group started working to formulate the contents for growth of bacteria, with the criterion that the artificial components used should exactly resemble the natural environment for microbial growth. Initially, he adopted the meat extract or infusion for culture, finally resulting in the nutrient broth or agar we use today. Nutrient Agar comprises 0.5% peptone, which is the enzymatic digest of meat, 0.3% meat extract (concentrated component of meat) and NaCl 0.8% (a concentration equal to concentration of salt, which is found in tissues). For fastidious organisms, the media was supplemented with vitamin, sugar, blood or serum. Further, Dr. Koch tried to convert the liquid media to a firm and rigid form, and for this purpose, he used gelatin in the broth.2 to 3% of gelatin was found to be suitable for this purpose because of its acidic nature, which had to be neutralized by adding sodium or potassium carbonate. The mixture was heated, forming a precipitate that was filtered to remove impurities (Koch, 1818).Then, a heat-sterilized container was heated at 150°C and the filtrate was poured and boiled in it for a short time; increasing the time period for killing the spores had the negative impact of making the gelatin lose its ability of solidification. Hence, it was concluded that normal wet heat sterilization could not be used for sterilization of medium containing gelatin as a solidifying agent (Koch, 1881).

Angelina Fannie and Walther Hesse of Koch's laboratory used Agar, a polysaccharide from red algae, as the solidifying agent to make solid media for growing microbes. Now gelatin was replaced by agar as a solidifying agent.

Walther Hesse was a community health physician and was a student of Max von Pettenkofer, who is today respected as the Father of Hygiene; after his military services, he joined Koch's laboratory and started working on microbial contamination of air and its quality. He also carried out investigations on water quality and developed techniques for bacteriological water testing. He also developed a culture medium for tubercle bacilli. His wife Angelina Fanny Hesse also assisted him in the capacity of a medical technologist apart from her household duties. While working on microbial contamination of air, he faced common problems of liquefaction of culture medium due to temperature constraint and sometimes bacterial strains were found to cause gelatin liquefaction. He was eager to find a solution for this problem and asked his wife about jellies and puddings remaining solid at the same temperature. As a youngster, Lina had heard about agar in New York from her Dutch neighbor who was from Java where agar had been used as a gelling agent in warm climates. The gelling property of agar, its thermal stability, resistance to microbial attack and long shelf life made it a suitable solidifying agent. This idea created a revolution in the field of microbiology, enabling the researchers to successfully conduct the investigations on tuberculosis. He immediately reported this to Koch; however, Koch did not highlight the contribution of Walther Hesse or his wife, in this regard Angelina Fanny also maintained silence about her contribution in the discovery because she was highly dedicated to her husband.

Agar has very specific properties because it melts at 100°C and remains solid even at 45°C (Walter and Angelina, 1882; Hitchins et al. 1939). Charles Chamberland in 1884 made a device for sterilization known as the autoclave, which is still used today for sterilization using the wet heat sterilization method. The word autoclave has a Greek origin, with the word 'auto' meaning self and 'clavis' meaning key, which translates into 'self-locking' device (Chamberland, 1884).In 1887, Sir R. J. Petri designed a special horizontal utensil, using double dishes of 10-11 cm in diameter and 1-1.5 cm height. Between the two dishes, one was of slightly larger diameter. These plates were sterilized by dry heat and after cooling, the lid was slightly opened and the media was poured into the smaller, bottom plate. Under these conditions, air-borne infection was found to occur rarely (Petri, 1887).

A technique for anaerobic culture was developed by Hungate (1950), who named it the roll tube culture technique. The techniques for bacterial growth and measurement of growth were given by Schaechter et al. (1958). This was correlated with the measurement of growth rate. Woese (1977) used 16S ribosomal DNA for identification of bacteria and identified Archaea having distinct genetic makeup; it was found to be related to both bacteria and Eukarya.

Anthrax was a fatal disease of domestic animals and R. Koch was investigating the causal organism of this disease. Before his work, it was reported that a rod-shaped bacterium existed in the blood of infected individuals and it could be transmitted either by dried or fresh blood to the healthy host. He was the first scientist to use mice as experimental animals because sheep and other animals were too costly. Koch obtained blood from Anthrax-infected animals and injected the same into the body of healthy mice, which died. He repeated the experiment twenty times and found reproducible results that spleen and blood were the targets of pathogen attack. Dr. Koch attempted to culture the pathogen, but did not succeed initially. Finally, a drop of aqueous humor from the eye of a cow was placed on cover slip and this was inoculated with a piece of spleen from the infected host. This cover slip was inverted on to a cavity slide for the hanging drop experiment. He observed the growth of pathogen with the formation of refractile bodies (spores). These spores were again inoculated to aqueous humor, yielding vegetative cells (Koch, 1882). Koch (1882) presented his paper on the etiology of tuberculosis in Berlin and confirmed the infectious nature of the disease. He also developed the methods for staining pathogens and preparing their culture, and suggested that infection was air-borne.

He confirmed the idea of his teacher Jakob Henle (1840) to prove the infectious nature of disease. The idea was that a pathogen (germ) was responsible for a particular disease and this pathogen could be used to form a pure culture; when this was re-inoculated into a healthy host, identical symptoms were produced. This concept was termed as Koch's postulates. A young medical student of Koch described this as a magical moment and considered this as a milestone because tuberculosis was widely spread in Europe and accounted for 300 deaths in a population of one million. However, the infectious nature of disease was not clear. This lecture was so impressive that millions of workers started working in the same area (Koch, 1882, 1884).

Ehrlich P used the dye methylene blue for staining the causal organism of tuberculosis, i.e., *Bacillus tubercule*. The dye was bound to the bacterium and this binding remained unaffected even after an acid and alcohol wash. Dr. Koch used an alkaline solution of methylene blue and after 24 hrs, it was treated with visuvin. The bacilli become blue while the background was stained brown. Methylene blue was used in an alkaline solution by Robert Koch. Hence, another alkali was used as a substitute, and aniline was found to be the most suitable. Sputum sample was taken between 2 cover slips, which were pulled apart from each other (Ehrlich, 1881).

The technique used by them was excellent, but there was the further need to differentiate between two bacteria. Around 1884, Christian Hans Gram, an associate of Herr Dr. Friedlander, began working in the morgue of the city hospital of Berlin on the cadavers of pneumonia patients, aiming to observe cocci in their lungs. This was used to observe the location of pathogens where nucleus and other tissues did not take up the stain while cocci were stained. He used aniline-gentian violet solution earlier used by Ehrlick, wherein the samples should remain in contact with the dye for 1-3 minutes, followed by contact with iodine-potassium iodide solution for 1-3 minutes. The samples turned dark blue violet in color. This color was not soluble in water, but in alcohol, these become clear. The samples were immersed in clove oil and it was observed that the remaining schizomycetes exhibited a vivid blue color.

Once the microbial culture technique was introduced, it was easy to identify various pathogens causing the disease (Gram, 1884).Most of the pathogens responsible for causing infectious diseases were identified in between 1877 to 1898 (Staley et al. 2007), as enlisted in Table 1.

J Lister, a British surgeon found that the death of patients undergoing surgery was more due to secondary infections and not due to surgery. He started investigating the factors involved and in 1870, started working on protection techniques for secondary infections. During his study, he found that patients with open wounds were more prone to infections in comparison to those who did not. During his investigations, he went through the work of Louis Pasteur regarding fermentation, vaccination and pasteurization and was also of the opinion that germs came from air (Krasner, 1995). He opined that the surgical instruments could be the source of the infections, and hence, tried several disinfectants for the surgical scissors. He used carbolic acid, and found it to be safe for human use. He highlighted the importance of hand wash, wound cover and sterilization of surgical instruments (Koch, 1881; Madigan et al. 2012).

Table 1. List of diseases and their Discoverers

Disease	Discoverer	Year	Causal Organism
Anthrax	Robert Koch	1877	*Bacillus anthracis*
Tuberculosis	Robert Koch	1882	*Mycobacteriumtuberculosis*
Cholera	Robert Koch	1884	*Vibrio cholerae*
Pyrogenic Infection	Alexender Ogston	1879	*Staphylococcus aureus*
Gonorrhea	Albert Neisser	1879	*Neisseria gonorrhoeae*
Diphtheria	Theodor Klebs	1883	*Corynebacterium diptherae*
Tetanus	Arthur Nicolaier	1884	*Clostridium tetani*
Typhoid	George Gafsky	1884	*Salmonella typhi*
Brucellosis	David Bruce	1887	*Brucella melitensis*
Gastroenteritis	August Gaertner	1882	*Salmonella enteritidis*
Bubonic Plague	Yersin & Kitasato	1894	*Yersinia pestis*
Botulism	Emile van Ermengem	1897	*Clostridium botulinum*
Dysentery	Kiyoshi Shiga	1898	*Shigella dysentrae*
Gangrene	Welch & Nuttall	1892	*Clostridium perfringens*
Sporotrichosis	Schenck	1898	*Sporotrichum schenckii*
Leishmaniasis DumDum Fever	Leishman Donovan	1903	*Leishmania donovani*
Syphilis	Schaudinn & Hoffman	1905	*Treponema pallidum*

Koch (1881) started working on the efficacy of chemical disinfectants and reported that carbolic acid was bacteriostatic, not bactericidal. According to him, concentration and contact time of the disinfectant affected its efficacy. He took spores of anthrax attached to a silk thread and exposed it to for different time periods to carbolic acid at different concentrations; each combination of time period and concentration yielded different results (Staley et al. 2007; Madigan et al. 2012).

Hensen (1873) found rod-like bodies when the edge of a leprosy nodule was cut with a knife and rubbed onto a glass slide. This was identified as *Mycobacterium leprae* and demonstrated it to be the causative agent of leprosy. Timothy Lewis (1878) started studying about parasites of Indian rats, and reported for the first time the existence of *Trypanosoma lewisi* as the causal organism of African sleeping sickness.

Malaria was another problem challenging the medical community regarding its agent and mode of infection. Laveran (1880) found that the malarial parasite infected erythrocytes. A bacterium was isolated from the human gut by Escherich (1885) and some strains of the bacterium were found to cause diarrhea and gastroenteritis; it was named as *Escherichia coli*. First report of a zoonotic disease came from the investigation of Smith and Kilbourne (1893), in the form of babesiosis in animals caused by *Babesia microti*, which is carried by ticks. Pfeiffer (1894) found that the cells of the cholera causing agent, *Vibrio cholerae* released a heat-stable toxic material on disintegration, and used the term endotoxin to describe it as it was bound to the membrane. The role of tse-tse fly and *Trypanosoma* was reported by Bruce (1895). Dr. Ronald Ross (1899) reported that malaria was transmitted by the bite of mosquitoes. The pathogen completed its life cycle in two hosts: man and mosquito. Walter Reed (1900) found that yellow fever was caused by a filterable virus and like malaria, it was transmitted by mosquitoes. This filterable agent was similar to that which caused foot and mouth disease of cattle. This was how a disease causing agent other than bacteria and fungi was reported. In the same year, Ophuls and Moffitt (1900) reported a new pathogenic mold *Coccidio desimmitis* as the causal agent of coccidioidomycosis. Darling (1906) observed a disease similar to tuberculosis, with the causal organism resembling *Leishmania* obtained from the patient's autopsy; it was ascertained to be *Histoplasma capsulatum*. It was thought to be a Protozoan. Ricketts (1909) working on Rocky Mountain spotted fever observed a causal agent that appeared to be an intermediate between virus and bacterium, and Rickettsia was, thus, discovered. Typhus, a vector-transmitted disease, was evaluated and found to be transmitted through body louse (Nicolle, 1910). The American microbiologist Alice Evans investigated Malta fever and its relation to unpasteurized milk. The causal organism was named as *Bacillus abortus* as it was known to cause miscarriage in animals and human. The disease was referred by her as Brucellosis. She was of the opinion that milk should be pasteurized to protect consumers from infections. Everitt Murray (1926) working as a microbiologist in Cambridge observed that a bacterium named *Bacterium monocytogenes* caused infection among laboratory rabbits. The bacteria was renamed as *Leisteria monocytogenes* in 1940, in the honor of J Lister. Riley and his colleagues (1959) proved that tuberculosis was an infectious disease and was disseminated as an aerosol. His results were based on experiments carried out with guinea pigs exposed to air coming from patients infected with pulmonary tuberculosis. A new class of infectious agent, prions was reported and it was found to be infectious protein material that caused scrapie disease, a neurodegenerative disorder. Montagnier and his colleagues (1983) first reported the HIV virus as the causal agent of AIDS and Marshall (1984) found that stomach ulcers contained a bacterium *Helicobacter pylorae.* The bacterium was found to use a special mechanism for protection against the acidic environment of stomach, converting urea to ammonia and tackling the acidity in the stomach (Warren et al. 1983).

Once the germ theory of discasc and causal organisms of various diseases were known, the classification of diseases started. On the basis of origin, diseases can broadly be categorized into two types. Diseases of infectious origin can be traced to a causal organism, whereas the other category of diseases are either due to nutrient deficiency or contact with contaminating chemicals. Before the discovery of antibiotics, infectious diseases were spreading to the scale of epidemic, even pandemic. Time to time, different control mechanisms were established such as-

a) Control through inhibiting contact with pathogen, such as removal of flies to control cholera, mosquitoes for malaria or rats for plague; similarly, restricted intake of contaminated water for control of water-borne diseases. John Snow

(1853) suggested that cholera epidemic could be controlled by preventing the consumption of water contaminated with human fecal matter (Staley et al. 2007).

b) Immunization: The method involves administration of vaccines to the healthy host, aiding the immune system in combating the pathogen.
c) Control through therapeutic agent

IMMUNOLOGY AND VACCINATION

Earliest report of inoculation of small pox is from 1549, post which there was a rather long lag period in 1713. Smallpox inoculation was popular in the Ottoman Empire, where Lady Mary Montagu observed the positive effect of variolation on the native population. Positively impacted, she performed the technique on her own children. Smallpox, an infectious disease, is spread through exposure to the *Variola* virus and starts quite innocuously like a common cold. Then the victim generally experiences fever, lethargy, muscle pain, and headaches. After a few days, rashes start appearing on the face and skin, with sores forming inside the mouth, throat, and nose. There is the characteristic development of fluid-filled pustules that expand and cover large areas of skin. After the third week of infection, if the patient survives, scabs form and separate from the skin.

Smallpox was widespread in the 18thcentury in Europe, accounting for an estimated. Investigators observed that patients who survived after infection became resistant to infection. In China, the scratched material from pustules was used to inoculate and powdered material prepared from the pustules was blown in the nose of healthy patient to prevent infection.

In May 1796, Edward Jenner, an English physician and scientist noticed a young dairymaid, Sarah Nelms, having fresh cowpox lesions on hand. Jenner inoculated James Phipps from material taken from her lesions. A mild fever was observed in the child; he lost his appetite, but recovered after ten days. Later in July, Jenner re-inoculated the same boy with fresh smallpox samples. Disease did not develop in him. Finally, Jenner concluded that the complete. We know today that after infection by the cowpox virus, there is development of immunity, and the infected person develops the ability to distinguish the similar smallpox virus from other antigens, becoming able to defend itself. Jenner continued to inoculate children with cowpox virus and got reproducible results. He termed this procedure as *variolae vaccinae* (smallpox of the cow), termed today as vaccination. Although he was not aware at that time, Jenner had laid the foundation stone for what is today called live attenuated vaccines. Today, some of the most common vaccines such as measles, mumps and yellow fever use a similar approach. Viruses are attenuated by passing them through a foreign host such as another live animal or embryonated eggs. During this process, the next-generation viruses lose their infection causing ability. The immune system easily destroys these weaker, less virulent viruses, leading to long-term protection. Although he was not the first person who tried to defeat smallpox and the trouble certainly did not end with him, it was the first scientific attempt to control an infectious disease by the intentional use of vaccination (Stanier, 1987; Ketchum, 1988; Krasner, 1995; Staley et al. 2007). Later, the practice of smallpox vaccination spread across the globe and ultimately saved millions of lives. The World Health Assembly announced global eradication of smallpox in 1980. Jenner's work acted as the fundamental for scientists to set the stage for further exploration for bringing the world into the age of modern vaccine research. He may be considered as the father of Vaccination technology.

Method of producing vaccines was discovered by L Pasteur in 1880 where he used attenuation technique for producing vaccine against the causal agent of chicken cholera. Dr. Pasteur was working on chicken cholera and incidentally, a culture of bacteria was spoiled and rendered incapable of causing infection on experimental organisms. Even when fresh virulent strain of bacteria was injected after the injection of former old culture, the chicken remained unaffected and expressed only mild symptoms. The discovery was an example of serendipity. He went on vacation and instructed his student Chamberland to inoculate chickens. Incidentally, Chamberland, too, went on a vacation; resuming work later, he used a one-month old culture to inoculate the chicken; interestingly, these developed only mild symptoms and then recovered. Chamberland assumed that an error has occurred, but Pasteur understood the underlying principle. He then applied this principle for immunization and developed a vaccine against anthrax. He also produced vaccine for rabies for the first time by growing the virus in rabbit, the pathogen weakening was achieved by drying the affected nerve tissue (Krasner, 1995). Vaccine was first created by Emile Roux, a French doctor and a colleague of Pasteur. First human trial was conducted on a nine-year old boy, Joseph Meister, at Pasteur's personal risk. The boy had been mauled by a dog, and brought to the doctor by his mother. Post the injection, the boy was kept under observation for three weeks, at the end of which period he exhibited complete recovery. This type of vaccines are still used in some countries and are cheaper, but have low efficacy and some side effects.

In 1888, E Roux and Alexender Yersin found that Corynebacterium diptharae produces a toxin known as the diphtheria toxin. This toxin directly affects tissues and organs. Culture filtrate obtained from cells is capable of killing laboratory animals directly, thus suggesting the role of toxin produced from the bacteria to establish the symptoms of the disease. Charrin and Roger (1889) found that serum had the capacity to agglutinate bacteria. The culture technique for Clostridium tetani was invented by Kitasato (1889) on the basis of the heat-resistant property of spores. He heated mixed cultures at 80°C for 1hr and in anaerobic environment, cultured the same. The technique of anaerobic culture from pathological samples was unknown before this discovery. These organisms were incapable of survival if exposed to air. Emil von Behring and Shibasaburo Kitasato (1890) were working on diphtheria and tetanus for protection against these diseases. Immunization of animals for tetanus depends on the toxin produced by the causative organism and cell-free blood fluid. Similarly, Von Behring working on diphtheria found that blood produces components that neutralize the diphtheria toxin. They found that

i. Blood of immunized organism has the capacity to neutralize the toxin
ii. The property is also found in cell-free extracts or extravascular blood
iii. This property is stable in blood and can move through transfusion.
iv. The property to destroy the toxin does not exist in the blood of animal, but appears after the injection of tetanus toxin in the blood. This was the beginning of toxoid vaccine.

Phagocytosis was observed by E Haeckel in 1862. Metchnikoff was working on Starfish larvae and on a day when his family was out to enjoy the circus, he observed mobile cells of transparent starfish larva; immediately, a thought came to his mind that similar defense cells were probably being produced in humans and animals as a defense mechanism against invaders.

Later, Metchnikoff initiated studies on diseases in *Daphnia* (an aquatic planktonic crustacean commonly called as water flea), which is a suitable model organism for such research because of its small size and transparent body wherein the changes can be noticed very easily. A fungal disease was noticed while observing the infections of *Daphnia*. It was observed that during the establishment of infection, the whole body cavity was filled with fungal mycelia and the causal organism was *Monosporabi cuspidata*. Several stages of life cycle from conidia to ascospore were observed. The fungus could not be grown on artificial media such as meat extract or orange juice. It was observed that the spores of fungus penetrated into the body cavity and blood cavity of *Daphnia*. When it entered the blood cavity, certain cells of blood attached to it, although the number of these cells was variable and the area of attachment became highly inflamed. During this time, some of the blood cells united to form a giant cell. On the other hand, spores present in the intestine remained unaffected, with the digestive juice having no effect on the spores. After binding the cells, changes in the shape and structure of spores were observed, probably in response to some secretion; fungal cells also produced a secretion deleterious to blood cells. However, the spores were finally ingested by blood cells. Thus, a clear-cut combat was noticed, for the first time, between the fungus and the blood cells. The blood cell defense activity resembled the phagocytosis property exhibited by lower organisms like protozoans. This was identified as a part of the body's natural immune system (Metchnikoff, 1884).

Karl Landsteiner (1901) discovered that people had different types of blood cells and thus, different blood groups. His discovery led to blood transfusions between people with compatible blood groups (Schwarz et al. 2003).

Richet Charles (1913), during his studies involving dogs in 1902, observed an opposite effect of phylaxis. After an initial low dose of a substance, a new dose of the same substance some weeks later could produce a severe reaction. He called this phenomenon as 'anaphylaxis' which stands for 'an opposite to phylaxis.' These results had great implications on our understanding of hypersensitive reactions.

Jules Bordet (1919) reported some proteins that help the immunological action of antibodies, which are called as complement proteins. He also reported that complement proteins bind to antibody-antigen complexes irrespective of the type of antigen or antibodies involved. He developed the complement fixation test that was used in diagnostics for several decades; he was awarded with the Nobel Prize in Physiology and Medicine in 1919. He also discovered anaphylatoxin and conglutinin. He first reported *Bordetella pertussis* as the cause of whooping cough and demonstrated the presence of lysozyme in human milk. In 1930, Burnet discovered the existence of multiple strains of the poliomyelitis virus, essential knowledge for the production of vaccine. While visiting Sir Henry Dale in London in 1931, he observed the discovery of the influenza virus by Richard Shope and mastered the developing-chick-embryo technique for virus culture. Burnet and his colleagues were the first to make influenza virus vaccine using the egg technique. In 1936, they discovered that Q fever, an infection in slaughterhouse workers with a worldwide distribution in humans, cattle, and sheep, was caused by a rickettsial organism. They also discovered, in 1952, the virus that caused the brain disease Murray Valley encephalitis, ascertaining that it was carried by migrating birds from New Guinea and transmitted by mosquitoes to man. Burnet explored the field of immunology and theorized that our immune system protects us against attacks by microorganisms and rejects foreign tissue. Part of our immunity has a hereditary basis, but part of it is acquired and is not present in the fetus. Burnet theorized that the ability to distinguish between one's own and foreign tissue is acquired during the fetus stage. His findings stimulated worldwide research in autoimmune disease (Owen et al. 2013).

Max Theiler (1951) reported that when the yellow fever virus was transmitted from one mice to another, a weakened form of the virus was obtained that could make apes immune to the disease. In 1937, Max Theiler succeeded in obtaining an even weaker variant of the virus. This variant, 17D, came to be used as a human vaccine. On March 26, 1953, American medical researcher Dr. Jonas Salk announced the successful development of a vaccine against poliomyelitis virus, which causes the crippling disease of polio. Polio is a disease that has affected human beings throughout the recorded history. It attacks the nervous system and can cause varying degrees of paralysis. Salk conducted the first human trials on former polio patients and on himself and his family, and by 1953 was ready to announce his findings. This occurred on the CBS national radio network on the evening of March 25, and two days later, in an article published in the Journal of the American Medical Association. After the clinical trials on nearly two million American school children, in April 1955, the vaccine was announced to be safe and effective. This initiated a nationwide vaccination campaign against polio. In 1962, an oral vaccine developed by Albert Sabin became available, greatly facilitating the distribution of the polio vaccine.

In 1954, John F. Enders and Dr. Thomas C. Peebles collected blood samples from several ill students during a measles outbreak in Boston, Massachusetts. They wanted to isolate the measles virus in the student's blood and create a measles vaccine. They succeeded in isolating measles from 13-year-old David Edmonston's blood (Owen et al. 2013).

Histamines are involved in allergic reactions, and it can be difficult to moderate their effects. Daniel Bovet (1957) looked for substances that block histamines, and in 1937, he found the first antihistamine, which later led to other antihistamine formulations to relieve allergies.

The first mumps vaccine was developed by Maurice Hilleman in 1967. He isolated mumps virus from his daughter, Jeryl Lynn, when she was ill with mumps at age 5.Consequently, the vaccine virus strain is referred to as the "Jeryl Lynn strain." Later on in 1969, Hilleman developed the first rubella vaccine by using rubella virus obtained from two scientists of the Division of Biologics Standards, Paul Parkman and Harry Meyer. Both the mumps and rubella vaccines developed by Hilleman contained live, attenuated virus particles. Hilleman's mumps and rubella vaccines were then used in combination with the measles vaccine, in the form of the measles-mumps-rubella (MMR) vaccine, which was licensed in 1971. The rubella component of the vaccine was changed in 1979, but in the United States, the mumps and measles vaccine viruses have remained the same since 1971. Measles vaccine can also be used in combination with mumps, rubella and varicella (MMRV) (Hendriks and Blume 2013).

Rodney Porter and Gerald Edelmann (1972) separately researched the chemical composition of antibodies. Using various chemical methods, they split up antibodies into smaller parts and mapped their biological effects. In 1959, this allowed them to determine the structure of antibodies. Antibodies are made up of two smaller and two larger polypeptide chains that form a Y-shaped structure. Peter C. Doherty and Zinkernagel (1973) proved how the immune system recognizes virus-ridden cells. A kind of white blood cell, T-cells, kills the virus-infected cells, but only if they differentiate both the foreign substances, viruses, and certain substances from the body's own cells. The discovery has not only provided an important basis for vaccines and medicines for infectious diseases but also for inflammatory diseases and cancer (Owen et al. 2013).

Cesar Milstein and Georges E. Köhler (1984) generated a hybridoma with the immortality of the myeloma cell and the ability to produce antibodies specific for sheep red blood cells, which originated from the spleen cell. After overcoming some technical difficulties, they successfully generated several hybridomas that secreted very large quantities of monoclonal antibodies. However, it took a couple of years for the broader implications of the discovery to be realized, when the potential advantages of the technique were outlined in an editorial in the Lancet in June 1977, claiming that the discovery could have profound implications for medical practice.

Tonegawa (1987) working on antibody diversity reported gene rearrangement in antibody production. This was reported to be possible through the redistribution of genes in a cell during its development into an antibody-producing B lymphocyte. In 2002, Sydney Brenner working on the small roundworm *Caenorhabditis elegans* gave the concept of programmed cell death. Brenner's work made it possible to link genetic analysis to cell division and organ formation (Owen et al. 2013).

Haraldzur Hausen (2008) reported that cervical cancer in humans was caused by certain types of papilloma viruses (wart viruses), the genes from which were incorporated into the host cells' DNA. This discovery made it possible to develop a vaccine against cervical cancer, which had been the second most common tumor disease in women (Owen et al. 2013).

Warfare against Microbes and the Discovery of Antibiotics

Paul Ehrlich, a co-worker in Koch's laboratory, suggested that a toxic dye might destroy bacteria without significant damage to the host and opined that organic arsenicals might be synthesized to eliminate invading parasites without hampering the host. Arsenicals (organic derivatives of arsenic- a toxic element synthesized and used on Trypanosomal infections on rats). The 606th compound he synthesized was found to be effective against Syphilis caused by a Spirocheate; it was termed as the Magic Bullet (Ehrlich, 1885; 1912).

Gerhard J. Domagk (1935) synthesized prontosil, which inhibited the growth of *Streptococcus*. The active moiety of the compound was sulfanilamide. He was basically an army doctor who later joined the department of sanitary services. During his services, he observed patients suffering from cholera, typhus, diarrhea and other infectious diseases and found himself helpless to control these infections.

During his investigations, he found a red dye and called it Prontosil, which when used on experimental organisms such as mice and rabbit provided protection against lethal doses of *Staphylococcus* and hemolytic *Streptococci*. This dye was a derivative of sulfanilamide (p-aminobenzene sulfonamide) synthesized by Gelmo in 1908; however, he was not satisfied with its action on human beings. The first patient on whom he experimented was his daughter suffering from a Streptococcal infection. Woods (1940) suggested the role of p-aminobenzoic acid in the mechanism of action of sulfanilamide used as a curative agent for Streptococal infections of mice.

Alexender Flemming was a professor of Bacteriology at St. Mary's Hospital Medical School in London. He was working on *Staphylococcus aureus*. In 1922 he discovered lysozyme, a component present in various biological materials such as egg white, tear, and mucus, which possessed strong antibacterial activity as it could cause the lysis of bacteria. The first biological material reported to possess antibacterial activity was obtained from a drop of his running nose that fell on his plate. Later he was working on the contamination of plates by airborne pathogens. His usual disorderly way of stacking his lab equipment actually led to a significant discovery. After a vacation, he observed that many of his plates were contaminated with fungus and he threw the plates in Lysol. A former member of the lab who was visiting the laboratory observed one of the discarded plates and found one contaminated plate showing a zone of inhibition around the fungus. Dr. Flemming realized the importance of this serendipitous discovery; he obtained the extract of the fungus and named it Penicillin (Flemming, 1929).

Florey was born in Australia and Chain in Germany, and both were influenced by the discovery of lysozyme and penicillin. They tried for large-scale production of penicillin, but only succeeded in producing penicillin in small quantities by using liquid broth cultures. They took eight mice and injected a lethal dose of bacteria into them; later, four of them were further injected with penicillin. Surprisingly, the latter four were found to be protected from infection. (Chain et al. 1945).

Wakesman was born in a small town of Ukraine, later moving to the USA for education. During his studies, he was given a research assignment for the assay of bacteria obtained from soil samples. He found a specific group of bacteria, filamentous in nature, which was closely related to bacteria in physiology and size but possessed similar structure to that of fungi. He further studied their role in soil processes such as decomposition of plant and animal parts as well as formation of humus. He observed their antagonistic effect on bacteria and fungus. Then he learned different techniques from his visits to France, Italy and Germany. He published a book titled 'Principles of Soil Microbiology' in 1927 and elaborated upon the metabolic aspects of microbes.

During World War II, he observed a need for new agents to combat infectious diseases. One of his students, René Dubos, isolated tyrothricin, which was effective against pathogenic bacteria. The role of soil in causing infections was also recorded by pioneer workers working on infectious diseases during this time.

He focused on Actinomycetes for the discovery of compounds exhibiting bioactivity against pathogens. He isolated thousands of microbes from soil and observed their antibacterial activity. A very small percentage of them were found to yield positive results for the activity; among this lot, further screening was carried out to isolate strains capable of producing the antibacterial substance in large amounts. Nearly 20 natural inhibitory compounds were observed, mostly from actinomycetes. He was the first one to coin the term 'antibiotic' for them. First compound of this group was actinomycin, as reported by Boyd Woodruff, a graduate student. It was effective against tuberculosis but was found to exert a toxic effect on human beings.

Woodruff then isolated Streptothricin that was active against Gram positive and Gram negative bacteria; although there was no immediate toxic effect on animals, delayed toxic reactions were observed. Breakthrough came with the discovery of *Streptomyces griseus* in 1943 by Albert Schatz. The compound showed broad-range activity against Gram positive and Gram negative bacteria, while penicillin was found to be effective against Gram positive strains only. Laboratory/clinical trials were carried out and streptomycin was found to be an effective chemotherapeutic agent against typhoid, tuberculosis, cholera, bubonic plague, and urinary tract infections. He provided the earliest definition of antibiotics. His autobiography was titled 'Life with the Microbes' wherein he wrote about constructive and destructive capabilities of microbes (Schatz et al. 1944).

Table 2. List of organisms and antibiotic activity

Antibiotic	Year	Discoverer	Organism	Site of action	Treatment
Doxycyclin (semisynthetic)	1940	Duggar	*S. aureofaciens*	Inhibitor of amino acid addition in translation	Pneumonia, acne, Prostatitis, Gonorrhea
Chloramphenicol	1947	Gottliab	*Streptomyces venezuelae*	translation	Typhoid
Neomycin	1949	Wakesman his student Huber Hubert Lechevalier	*Streptomycec fradiae*	translation	Tuberculosis & it has side effects used in skin infections
Oxytetracyclin	1950	Finlay and group	*Streptomyces rimosus*	translation	Spirochete, Clostridial infection, anthrax, respiratory, urinary tract infection, skin, ear, eye and gonorrhea
Erythromycin	1952	Abelardo-A guilar- Eli-Lilly company	*Streptomyces erythreus*	translation	Respiratory, dental infections but have some side effects.
Vancomycin	1955	Dr. E. C. Kornfeld, Eli-Lilly	*Streptomyces orientalis.*	Cell wall	Gram Negative bacterial infections
Rifampicin (semisynthetic)	1957	Piero Sensi and Maria	*Amycolatopsis rifamycinia*	Transcription	Tuberculosis, leprosy
Gentamycin	1963	Marvin Weinstein	*Micromonospor a purpurea*	Translation	Bone infection, Meningitis, Pneumonia, Urinary tract infection
Imipenem	1970	Burton and group fromMerk	*S. cattleya*	Cell wall	Effective against both Gram Positive and Negative bacteria

The discovery of antibiotics proved to be a miracle and caused a revolution in medical sciences, overcoming many a disease incidence. With the discovery of Penicillin and Streptomycin, effective treatment was now available against diseases, and the same were routinely used for every infectious disease. However, this led to the development of resistance against pathogens, and the new challenge now was to discover other antibiotic drugs. Aureomycin was the first tetracycline discovered in 1948 by the American Botanist Benjamin Minge Duggar, University of Missouri, Washington. Later, he was working as a consultant in the American Cyanamid Company, and noticed that penicillin and streptomycin were not found to be effective due to development of resistance. He started working on the isolation of strains from soil capable of producing antibiotics. During his investigation, he analyzed 3500 strains of mold from soil and later found one interesting sample from the University of Missouri campus. A golden yellow-colored substance produced by mold exhibited antibiotic property, which was named Aureomycin (Latin *aureus* means gold; *mykes* means fungus). The antibiotic was effective against Rocky Mountain spotted fever, Rickettsia, Trachoma, parrot fever, typhus, chlamydias and mycoplasma. List of some antibiotics have been given in Table 2.

Antibiotic discovery program will continue until the existence of life on the earth, because of the existence of resistant strains and emerging infections.

THE ORIGIN OF VIROLOGY

Bacteria and fungi were reported as pathogens causing diseases in animals and human beings, but virus, another common infectious agent was not reported in the early years of microbiology. Iwanovsky (1892), who was working on plant diseases, may be credited as the one to initiate virus-related studies. He observed the tobacco mosaic disease, which was thought to be caused by a bacterium species. He used the filter method for isolation of bacteria and found that sap from the diseased plant could infect the healthy host. He found that this agent was extremely small and not visible even at immense magnification. The organism could penetrate porcelain filters that were known to retain bacteria. Although Pasteur developed the vaccine for rabies, he was not aware of the causal organism of the disease. He was of the opinion that the pathogen could not be detected by microscope because of its small size; it could even cross the Chamberland filter with a pore size smaller than bacterial size. Working before Iwanovsky, Mayer (1876) had observed that TMV was infectious and the disease was caused either by a toxin or by a very small agent. Beijerinck (1899) found that after filtering the sap from an infected plant, a new form of infectious agent was noticed. The agent exhibited the capacity of multiplication only when present in the host. He used the term ‘contagium vivum fluidum’ (soluble living germ) and coined the term virus. He was of the opinion that organisms exist in the liquid stage. Stanley (1935) isolated a crystalline material from

Turkish tobacco plant infected with virus and this crystalline material showed similarities with the tobacco mosaic virus. The crystallized material was found to contain 20% nitrogen. The crystal obtained had capacity to infect tobacco plants. It was later found that in addition to protein, the crystalline material also contained ribose nucleic acid. In comparison to the plant viruses, animal viruses were found to be more complex. Bacteriophages are the viruses that infect bacteria, Twort (1915) may be considered the first researcher to work on them. Unfortunately, his work was not noticed and later D'Herelle (1917) found an invisible agent having antagonistic activity against bacillus of Shiga (dysentery bacillus). He used broth seeded with few drops of feces from patient suffering from dysentery, and the cultures were incubated. After full growth, the contents were filtered through Chamberland candle filter. A small amount of active filtrate was added to the broth culture of Shiga. Growth of culture was inhibited with complete lysis of cells within hours and days. The filtrate when again applied to fresh bacterial cultures, caused lysis. The parasitism was highly specific for specific bacteria. Later, viruses acting against cyanobacteria was reported by Safferman and Morris (1963) and named cyanophages (LPP1).

The viral disease causing agents were distinguished from bacteria and this led to the establishment of virology as separate branch (Rivers, 1926). The size of virus was elucidated and it was found to range between sizes of large protein to tiny bacteria (Elford, 1931). At that time any technique was not available for culturing of viruses and this technique was reported to use eggs for cultivating them (Woodruff and Good Pasteur, 1931). Although organisms like bacteria were visible under microscopes with high resolving power, but no instrument was available to observe small organisms like viruses. Ruska (1931) designed and fabricated the first electron microscope, which was finally modified by Knoll in 1931. Hey used electron microscopy for taking the picture of Viruses. Luria and Anderson (1942) used this microscope for identification and characterization of Bacteriophage and discovered a head and tail structure in phage.

INDUSTRIAL MICROBIOLOGY

Industrial microbiology has been known from the ancient times. Indeed, one can find evidence of beer and wine making in 7000 BC (Assyria, Mesopotamia), 6000 BC wine making (Georgia), 3000 BC fermented milk products (Babylon), 2600 BC bread (Egypt), 1000 BC soy sauce (China), 600 BC cheese (Asia) and since time immemorial in India.

The process behind the industrial production of valuable compounds is fermentation. The word fermentation is derived from the Latin word *fervere,* which means to boil; it is related to the bubbles observed in a fermented solution. It is a process where an organic compound acts both as an electron donor and an electron acceptor. For a microbiologist, fermentation is the process by which a wide range of products – from alcohol to single cell protein is produced by through microbial processes, through aerobic or anaerobic processes. The transformation of food to delicious fermented foods was thought to be a miracle and at one point of time, it was even thought to be the fruit of divine powers. Osiris was thought to be the driving force leading to richly brewed beer and Greeks believed Bacchus to be the god of wine. The existence of microbes was not established in fermentation. During this period, two of the major challenges facing scientists was about the nature of fermentation and the nature of enzymes. In 1700, Lavoisier studied the transformation of sugar to alcohol and CO_2 and suggested that the amount of sugar consumed was equal to the amount of alcohol and CO_2. The idea was also confirmed by Gay Lussac, and the entire process was thought to be a purely chemical one. George Stahl (1697) was of the idea that yeast played a physical role in the conversion rather than chemical. The credit for recognizing the role of microbes in fermentation should be given to Cagniard de la tour, Schwann and Kuetzing, who found a living organism capable of reproduction by budding. Various chemists of that time were opposed to the view that living systems were involved in fermentation. Finally, it was Louise Pasteur who proved the biological nature of fermentation (Krasner, 1995). He proved that lactic acid fermentation was caused by a living organism. He defined the fermentation process as life without air, and was of the opinion that different microbes produced different end products. Till now, the relevant fermentation enzymes were not discovered; it was only in 1897 that the German chemist Eduard Buechner found that dead, grounded yeast when added to sugar solution could still ferment it. Finally, it was established that the fermentation process was carried out by enzymes produced by microbes (Staley et al. 2007; Madigan et al. 2012).

Production of butanol was reported by Pasteur (1861) and of acetone by Schardinger in 1905. Fernbach developed a technique for bacterial fermentation for butanol production using potato starch as substrate. ABE (Acetone, Butanol and erthanol) fermentation was established in 1916 by Weizmann from *Clostridium acetobutylicum,* who laid the foundation of modern industrial microbiology. Another important product was citric acid, which has wide applications and was discovered by Scheele in 1784. Currie (1917) found *Asperjillus niger* to be an efficient producer of citric acid, leading to the industrial production of citric acid by Pfizer. Lactobacillus producing lactic acid was discovered by Pasteur (1856).

Another important industrial product of microbial origin with high commercial value was amino acids, which are used as animal feed additives (Lys, Met, Thr), flavor enhancers (monosodium glutamate, Ser, Asp) and pharmaceutically important compounds. Traditionally, kelps were used as flavor enhancers. However, K Ikeda of the Tokyo University isolated a flavor enhancer from *Laminaria japonica* and named it monosodium glutamate. Kinoshita, S working at Kyowa Hakko Kogyo Co. isolated a soil bacterium producing high amounts of glutamic acid. The isolation technique was interesting, where the producer was grown in duplicates and one set was treated with UV light to kill colonies; now, this plate was over-laid by *Leuconostoc mesenteroides,* a bacterium fastidious for glutamic acid. The lower layer of colonies excreted glutamic acid and supported the growth of *Leuconostoc*. He isolated the bacterium *Corynebacterium glutamicum* as a commercially significant producer of glutamic acid (Glazer and Nikado, 1995).

Another important microbial product of high economic value is microbial enzymes. Takamine suggested the technical possibility of cultivated enzymes, and their use in industry. Biodin and Effront pioneered the production of enzymes in 1917 (Glazer and Nikado, 1995). Enzymes are also known to play an important role in the production of therapeutically valuable steroids, such as cortisone, and this property of biocatalysis was already used 50 years ago. Microbes are capable of catalyzing various reactions such as oxidation and reduction of steroid and sterol substrates, which is an excellent manifestation of their regio-selective and stereospecific biotransformation property. Biotransformation is the property of biological agents to convert one compound to another. This property of microbes was exploited for hydroxylating any position in the carbon skeleton of a steroid nucleus by employing the enzymes present in them. These enzymes are steroid hydroxylases, which are named on the basis of the position they attack on the rings or the side chain of the steroid nucleus. Once the property of microbes was established, it dramatically lowered the cost of production of steroids and antibiotics. In the year 1930, Edward C. Kendall and Tadeus Reichstein from the Mayo Foundation and the University of Basel, respectively, together succeeded in isolating cortisone for the first time from the adrenal gland. This was found to be effective in patients suffering from acute rheumatoid arthritis (Philip S. Hench, 1949) because the compound had anti-inflammatory effects (Kendall, Reichstein, and Hench were awarded Nobel Prize in 1950). This immediately resulted in a high demand of cortisone, but the only route of synthesis was chemical transformation, which was a 31-step long process with low yield. The precursor of cortisone was deoxycholic acid, which was obtained from beef bile, although the yield was low with the market price of $200 per gram. The most difficult challenge here was the shifting of hydroxyl group from carbon 12 to carbon 11, which was a 9-step process. Later, in 1952, researchers at the Upjohn Company discovered that the aerobic bread mold *Rhizopus arrhizus* could hydroxylate progesterone (early intermediate in cortisone synthesis) at the 11^{th}carbon. Later, another common mold *Aspergillus niger*, was found to carry out the same reaction by researchers at the Squibb Institute. This reduced the number of transformation steps from 31 11; another benefit of using microbial catalyzed reaction was its eco-friendly nature. Most important factor was the commercial feasibility of the process, reducing the price of the product from $200 per gram to $6 per grams. Further reduction in price came through the introduction of an alternate precursor for synthesis; this was diosgenin, which was obtained from the Mexican barbasco plant. Here, side chain beyond carbon 21 was removed by using mycobacteria. Now, the cost of the product came down to 46 cents per gram by 1980.

Later, steroid nucleus was further modified with the help of microbes to produce cortisone, aldosterone, prednisone, and prednisolone.

These days, even such products are being microbiologically manufactured wherein microbes do not generate the product naturally – this has been made possible by using recombinant DNA technology such as recombinant vaccines, interferons, insulin and blood clotting factor (Glazer and Nikado, 1995).

Environmental Microbiology

Microbes play important roles in environmental processes, but the role of microbes in the environmental processes were not reported until recently. Ferdinand Cohn (1872) was a pioneer in this field and was of the opinion that microbes played an important role in the cycling of nutrients. The term *Bacillus* was coined by him to describe the rod-shaped bacteria. Winogradsky carried out the first careful observation on the role of controlled environment on the growth of microbes in pure culture (Wakesmann, 1954). Winogradasky started working on *Beggiatoa,* which had already been studied, and found it to accumulate sulfur granules (Cramer, 1870). Cohn (1875) also found similar accumulation of sulfur granules in bacteria residing in sulfur springs. He collected samples of such bacteria from natural environment and started culturing them. He manipulated culture conditions by placing the bacteria in Geisser chamber containing growth media. He observed that bacterium isolated from natural habitats had sulfur granules. Cohn found that hot water springs were saturated with H_2S. A question that faced Winogradasky was whether sulfur granules were produced from the reduction of sulfate or the oxidation of H_2S. In the growth chamber, he used spring water containing H_2S and sulfates. He observed that the presence of H_2S resulted in the formation of granules while $CaSO_4$ caused the death of organism. In another experiment, he used slide culture technique where bacterial suspension was placed beneath the cover slip and observed that these aggregated to forma ball-like structure. When H_2S was introduced to the edge of the cover slip, the bacteria migrated (Winogradsky, 1887). In the absence of H_2S, S granules were not observed. He reported that after oxidation of H_2S, sulfur was further oxidized to its highest oxidation state e. g. H_2SO_4 and introduced the concept of chemolithotrophy, where bacteria obtained energy by oxidation of reduced inorganic compounds such as H_2S, NH_4, Fe^{2+}.This suggested the role of microbes in sulfur and nitrogen cycles. In 1861, Pasteur found that ammonia is oxidized to nitrate through a biological process. Schloesing and Muntz (1877) found that when sewage containing ammonia was incubated there was the appearance of nitrate through a biological process. This formation of nitrate was inhibited in the presence of chloroform.

Louis Pasteur (1861) reported the existence of life in anaerobic environment and coined the term aerobes and anaerobes for microbes surviving in aerobic and anaerobic environments. He found that more alcohol is produced from sugar solution in the absence of oxygen (Krasner, 1995; Staley et al. 2007).

Hellriegel and Wilfarth (1988) reported, for the first time, the role of some organisms in symbiotic nitrogen fixation. Denitrifiers were isolated to pure culture by Gayon & Dupetit in 1986. Although the soil scientists were aware of the role of microbes in loss of nitrogen from soil, they were not able to identify the same. Reiset, (1856) was the first one to suggest that in the biological world, nitrogen is recycled.

McCrady (1911) invented a technique for the analysis of water samples to check for the presence of coliform bacteria. The technique is known as the Most Probable Number (MPN), multiple tube fermentation technique, which is popular till today. In this test, presence of coliform was detected on selective medium for their ability to grow and to produce acid or gas by utilizing a particular carbon source at a particular temperature. The number of coliform bacteria was measured by using the MPN tables.

Microbes display immense diversity for the utilization of different compounds as a source of carbon. Methane utilizing bacteria, the methanotrophs were observed by Sohngen (1906). He also observed a different class of bacteria capable of producing methane, and called them as methanogens. Bacterial photosynthesis was found to be different from the process observed in plants, as photosynthetic bacteria were found to use reduced compounds as electron donor; this was, indeed, anoxygenic photosynthesis. Sulfur bacteria were found to use H_2S as the electron donor while plants use water as the source of electrons and exhibit oxygenic photosynthesis (Van Neil, 1931).Later, Wood and Werkman (1936) found that *Propionibacteium arabinosum* used CO_2 during the fermentation of glycerol. This was the first report of CO_2 fixation by heterotrophic bacteria.

Fungi play an important role in the decomposition of organic matter and formation of humus, thus playing an important role in carbon cycling and increasing soil fertility (Koning, 1904).Quorum sensing is a phenomenon of auto-regulation in bacterial systems and plays an important role in the process of biofilm formation. Pathogenicity was observed in *Vibrio fischeri,* a light emitting bacteria, which produces a compound AHL that acts as an autoinducer, and its concentration depends on cell density. This was observed by Hasting and Gibson, (1967).

The upper limit of growth temperature tolerated by bacteria was thought to be 73°C. Brock (1967), while on a trip to the Yellow Stone National Park, observed a microbial mat and isolated a pink-colored bacterial sample growing at85°C. Its growth was also observed at this temperature. The observation was proved to be a milestone because it led to the discovery of an entirely new microbial domain, Archaea. Endosymbiotic hypothesis regarding the origin of mitochondria and chloroplast was given by Margulis (1966); according to the theory, mitochondria originated from aerobic bacteria while the latter could be traced to cyanobacteria. Adler (1969) conducted an experiment using a dense culture of bacteria; a chemical-containing tube (containing galactose, ribose, aspartate and serine) was inserted into this flask, and it was noticed that the cells accumulated towards the chemical. He concluded that bacteria had receptors for sensing changes in environmental conditions.

With the ever-increasing human population, there is continuous decrease in fertility of agricultural land and per unit productivity. This has led to food crisis in several parts of the world; deficiency diseases such as Kwashiorkor and Marasmus can be directly traced to the lack of access to nutritious food. Microbial cells are important in this context, as they are rich source of proteins. *Spirulina* was used in the Chad area of Mexico as dried cake and this was specially provided to pregnant women. They were of the belief that it would protect the newborn from the evil eye. Scientifically, though, this was due to the high protein content in the cells of *Spirulina*. Research on single cell protein technology started when Delbruck observed that Brewer's yeast could be used as a feed supplement for animals because of its high nutritive value (Ugalde and Castrillo, 2002).A process known as Zulaufverfahren, which means fed-batch, was reported by Sak in Denmark and Hayduck in Germany. In this process, diluted sugar solution was added continuously to an aerated suspension of yeast; this was in contrast to the earlier process wherein both were added at once (batch culture) (Ugalde and Castrillo, 2002). In 1960, a group of workers from the British Petroleum Industry used paraffin, a byproduct of the petroleum industry, for single cell production. The initial work in this regard had been established by Champagnat in France (Bamberg, 2002). The term single cell protein was coined by Wilson from the Massachusetts Institute of Technology (Dolle, 1994).With the emerging knowledge of microbiology, different new components were discovered to play important roles in normal life. For instance, biopolymers were discovered to overcome the problem of synthetic polymers, which do not degrade in nature. Tilden (1935) described various uses of agar, a polymer. Later, another polymer, xanthan gum was reported from a bacterium by Jeanes and her team, commercial production of which started for various uses under the trade name Kelzan. A French researcher Lemoigne (1926) reported a bioplastic material polyhydroxybutyrate in 1926, while working on *Bacillus megaterium*. This was used for the production of biopolymer higher tensile strength than synthetic polymer, and the added benefit of biodegradability (Glazer and Nikado, 1995)

AGRICULTURAL MICROBIOLOGY

Just like animals and human beings, plants also suffer from different diseases; in fact, causal organisms are also the same, including fungus, bacteria, mycoplasma, nematodes and viruses. The concept of plant diseases was known from ancient times, although it was not an established branch of science. Existence of diseases has been recorded in ancient literature such as the Vedas composed as early as 1200BC. The symptoms of plant diseases were described in Vrikshayurveda by Surapala in ancient India. Theophrastus (300BC), known as the Father of Botany, observed some diseases in plants and suggested few measures to control these.

Lord Pliny (100AD) also noticed the existence of diseases in plants, and he believed that environmental factors played an important role in the development of diseases. However, these concepts could not be developed because of a lack of facilities to study the causal organism. The world of microbes was yet to be discovered. After the discovery of microbes, Micheli (1729) discovered the morphology of fungi and was of the opinion that they originated from spores. He has been considered the Father of Mycology. Tillet (1755) proved that contagions causing human diseases were also involved in plant diseases. He proved that Bunt of wheat could be controlled by seed treatment. Like in the case of human diseases, any therapeutic agent was not known to conquer plant diseases. Prevost used $CuSO_4$ to control Bunt of wheat. Fontana & Tozzetti (1667) demonstrated for the first time that stem rust fungus was involved in the rust disease of wheat. Thirty years later, Persoon gave it the name *Puccinia gramins.* Louis Rene and Charles Tulasane discovered five spore stages in the life cycle of fungus.

Similar to the epidemics of human diseases such as plague, cholera, smallpox, and cholera, the potato famine of Ireland occurred between 1845 and 1857; potato crop was heavily damaged due to infection of fungus *Phytophthora infestans,* causing late blight of potato. The potato was a cheap crop and formed the most staple food of many. A loss of this scale caused a famine that resulted in the death of one million people and migration of about as many. It was Anton de Bary (1853) who proved the causal organism to be *Phytophthora infestans.* The name was given because of its nature –while `phyto' means plant, 'phthora' means eater. However, it was only after Koch's postulates had been laid that this fungus was confirmed as the exact causal organism. Burril (1878) was the first to report that bacteria could cause disease in plants. He found that pear and apple blight disease were caused by bacteria. France was the largest producer of wine, which is produced from grape. The grape crop was destroyed by mildew disease and in order to control this infection, Bordeaux mixture was discovered. The Bordeaux mixture was a mixture of copper sulfate and slaked lime.

Plant growth and productivity was related to nutrient status of soil and in order to increase productivity application of chemical fertilizers started. However, the use of chemical fertilizer was not eco-friendly. The scientific community started looking for a safer but equally beneficial approach. Theophrastus was the first to state that mixing of different soils increased agricultural productivity. The concept of biofertilizer came into existence. The concept was now understood that mixing of soils caused mixing of different microbes that are beneficial for agriculture. Although the concept of biofertilizer was known from a long time period, it was observed that crop rotation with leguminous crops enhanced productivity of the following crop. In 1888, Beijerinck was the first to isolate a microbe from the root nodule of a leguminous plant and termed it *Bacillus radicicola.* Biological nitrogen fixation was reported for the first time. In 1890, the anaerobic nitrogen fixer *Clostridium pasteurianum* was discovered by Winogradisky. This was the first free-living, non-symbiotic nitrogen fixer. In 1893, Noble and Hiltner introduced the nitrogen legume inoculum to crops. Soon, nitrogen fixing microbes were discovered. The first free-living aerobic nitrogen fixer *Azotobacter* was discovered by Beijerinck in 1901, while the micro aerobic nitrogen fixer *Azospirillum* was isolated by him in 1922 from garden soil, using nitrogen-deficient media. The cultures were found to increase nitrogen at the expense of malate. Cale Gessard (1884), a chemist and bacteriologist, isolated a microbe that produced water-soluble pigments, which when exposed to UV light turned blue green – this was the first report of *Pseudomonas.*

The role of microbes was established in growth promotion and their application was biofertilization. Later, Kloepper introduced the term Plant Growth promoting Rhizobacteria in 1970 (Kloepper and Schroth, 1978). The bacteria residing in the rhizospheric region promote the growth of plants and their effects on plant were reported in 1904 by Hiltner, who used the term 'rhizospheric effect'.

Insect pests damage an enormous amount of crops, resulting in immense loss of yield. In the mid-1800s, the concept of chemical control of insects was introduced by farmers. These chemicals were organic and inorganic man-made compounds such as organophospahates, organochlorine compounds, carbamates and pyrethroids etc. The major disadvantage of using these was the development of resistance and elimination of non-target pests. These compounds are non-biodegradable and toxic to humans, too. The scientific community started searching for safer alternatives. The concept of Biopesticide emerged, one of the approaches of which is to find out a microbial pathogen to infect a particular insect damaging the crop. Isiwata (1905) in Japan isolated *Bacillus thuriengiensis* from a diseased silk worm larvae and named it as *Bacillus sotto*. Berliner E isolated a similar pathogen from a diseased Mediterranean flour moth (*Anagasta kiibniella*) and gave it the name *Bacillus thuriengiensis* as it was the province Thuringen from where it was isolated. The knowledge remained unexplored up to 1976, till when pathogenic strains were known only against Lepidopterans. Dr. Tahori and Margalit (1975-76) were conducting a survey in Israel for potential biocontrol agents against mosquitoes; during sampling they found a small pond in a dried out river bed near north central Negev Desert near Kibbutz Zeelim. The water here was brackish and with its high load of organic matter, it was serving as a mosquito breeding site. In this mosquito breeding site, dense population of dead and dying larvae of *Culex pipiens* (a common species of mosquitoes) was noticed as a thick carpet on the surface. On the other hand, pupae and sunken adults were observed attempting to emerge from their pupal cases floating on the surface. Samples were collected from different sites, including the edge of the pool, the dead and decomposing larvae, the water and the silty mud. The samples were used to isolate bacteria in the lab, in association with Mr. L. H. Goldberg, and purified to single colonies. Single colony of the isolate was designated as ONR 60A, and found to be *B. thuringiensis* var. *israelensis*. Insecticidal preparations of *B. thuringiensis* strains were found to be toxic to mosquitoes and blackflies, and are now used successfully as biological control agents in various parts of the world. Biopesticide production is now a well-established industry (Glazer and Nikado, 1995).

BACTERIAL GENETICS

Microbes such as bacteria are prokaryotic in nature, containing a haploid number of chromosomes and lack reduction division as well as sexual reproduction. Therefore, the reason for development of recombinants was not known. First report of recombination in bacteria came through the work of Griffith (1928). He found two strains of bacterium *Streptococcus pneumonae*, one with a smooth capsule (S) and the other with a rough capsule (R). The former was virulent while the latter was found to be non-virulent when injected into experimental mice. When a combination of heat-killed S and R types were injected, virulence properties were developed in the R strain. He was of the opinion that some transforming principle moved from S to R and was responsible for the development of virulence. This was the first report of exchange of genetic material and was termed as transformation. This was also regarded as the foundation stone for the discovery of DNA as genetic material. Beadle & Tatum (1941) working on *Neurospora crassa*, an ascomyceteous fungi found that a particular gene was expressed by producing an enzyme. They concluded that a particular gene controlled the formation of an enzyme for the production of niacin. They gave the 'one gene one enzyme' concept. *Neurospora* growing on culture media containing nutrients essential for growth were exposed to X – Ray; some colonies developed while some failed to grow. These colonies could develop only after the addition of Vitamin B6. They found that the defect was inherited as Mendelian recessive genes. Mutation was reported by Hugo de Vries in 1901; regarding the appearance of spontaneous mutation, there was considerable debate pertaining to its origin. Luria and Delbruck (1943) were working on the appearance of resistance in bacteria against phage attack. They found that inheritance of characters in bacteria also followed Darwinian principles. Viral resistance could develop even in the absence of virus. They designed the fluctuation test and gave proof against the adaptation hypothesis of spontaneous mutation. Avery et al. (1944) extended the work of Griffith and used pancreatic deoxyribonuclease in the transforming experiment; however, transformation was not noticed when pancreatic ribonuclease and protease enzyme was used. Avery et al. (1944) gave the first indication of DNA being the genetic element, giving a new direction to research in molecular biology.

Another method of gene transfer and recombination in bacteria was observed when Lederberg and Tatum conducted an experiment with two auxotrophs unable to grow on minimal media when grown together, the appearance of prototrophs was observed. Based on this, they reported a process known as conjugation for the transfer of genes from one bacterium to another. When U tube was used to separate two auxotrophs separated by sintered glass filter, recovery of prototrophs were not observed. Thus, physical contact was necessary for the conjugation process (Lederberg and Tatum, 1946). In the early 1950s, during initial studies on *Bacillus megaterium,* Andre Lwoff and his colleagues at the Pasteur Institute described the phenomenon of lysogeny in *E. coli*. It was clear that some strains of *E coli* were lysogenized by phage, which meant that the bacteria harbored phage in a dormant form called as prophage. These bacteria grow normally and could not be easily recognized as lysogens.

Lwoff then exposed lysogens to a moderate dose of UV light; remarkably, the cells stopped growing and after 90 min, they were lysed to release a fresh crop of phage particles. The phages were not capable of infecting an already lysogenized bacterial cell. This was referred to as immunity to super infection. The experiment of Luria and Delbruck was further confirmed by Lederberg (1952) by conducting the replica plating experiment, which confirmed that spontaneous mutation was independent of adaptation. Existence of extra-chromosomal genetic element in bacteria was reported by Lederberg in 1952. The plasmids present in bacteria carried out functions other than the functions of genomic DNA. Final proof for DNA as genetic element came through the work of Hershey and Chase (1952), who found that radioactivity of phosphorus was noticed in the next generation of phage while radioactivity of sulfur remained unnoticed. Radioactive sulphur was a part of the protein coat while radioactive phosphorus was a part of the DNA, which was transferred from one generation to another.

Once conjugation was reported in bacteria, researchers began to form the opinion that in eukaryotic systems, the parallel phenomenon was that of fusion. According to Hayes (1952), conjugation in bacteria was the unidirectional transfer of genes from donor to recipient. He, along with Cavallis and Lederberg observed that the 'fertility factor' was present only in donor cells. Circular nature of bacterial chromosome was soon reported by Jacob and Wollman (1957), through an experiment called the 'interrupted mating technique.' Interestingly, antibiotic resistance can be transferred between *Shigella* strains and *E. Coli* strains and this phenomenon does not involve transduction and transformation (Ochiai et al. 1959). Paradee et al. (1959) working on Galactosidase observed that its expression was induced by changes in culture conditions. They were of the opinion that a repressor protein regulated the expression of gene by binding to a site on the DNA. Even up to this time, the role of an enzyme in DNA synthesis was not known. It was only later that an enzyme synthesizing DNA in bacterial cell-free extracts was observed (Kornberg, 1960).The enzyme was DNA polymerase, which needs a DNA template for the synthesis of a new strand. Jacob and Monod were part of a group working on lac operon; they isolated several mutants of *E. coli* and observed several sequences attached to structural genes to regulate their expression; others were working on a protein that works away from its site of origin. This led to the origin of the concept of cis-acting and trans-acting molecules. They provided the operon model of gene regulation. Once the lac operon was discovered, the tryptophan operon followed suit. It was found that the tryptophan operon worked on a different mechanism of gene regulation, not involving the repressor protein. The primary transcript had a region for the synthesis of a short peptide known, as the leader peptide. This region contains two codons for tryptophan and availability of this regulates the transcription process for the expression of structural genes; this process was termed as attenuation (Yanofsky et al. 1962).Bruce Ames (1966) designed a test for the determination of mutagenic potential of compounds, wherein he used auxotrophs and treated them with the chemical in question to observe their mutation to the corresponding prototroph. Smith et al. 1970 reported enzymes isolated from *Haemophilus inflenzae*, which cut DNA at specific sites. This was the very first report of restriction enzymes, which opened a new branch for manipulation of DNA through rDNA technology. Use of restriction endonuclease to cut DNA at predictable manner led to the formation of the first recombinant DNA molecule by the joining of viral and bacterial DNA (Jackson et al. 1972). The subsequent discovery of ligases – enzymes that join two strands of DNA – was the next link in the chain.

Temin was of the opinion that coat protein of certain viruses had an enzyme that caused duplication of viral genes. He suggested that genetic information flowed from RNA to DNA. Finally, it was considered as RNA dependent DNA polymerase (Temin and Baltimore, 1970).

In 1995, Venter's team sequenced the genome of *Mycoplasma genitalium*, which is a sexually transmitted bacteria having the smallest genome of any known free-living organism (Peterson et al. 1995). Craig Venter (2010) created a synthetic cell that contains the smallest genome of any known, independent organism with 473genes. His group inactivated genes one by one to test whether the bacterium could still survive, trimming down the list of this synthetic cell's genome size 375 genes. The cell contained only those genes that were essential for its survival. It can be grown in axenic cultures and is called as a minimal cell. This can serve as a model to yield fundamental knowledge of the essential genes, parts, organization and machinery of a living cell. In simplest terms, the minimal cell is the essence of a living system and raises stakes in race to harness synthetic life.

Conclusion

Microbes are important role players in our everyday life and are present everywhere on the earth. They play important beneficial roles in the preparation of food products, alcoholic beverages, pharmaceutical compounds, bio plastics, and energy sources. On the other hand, they are also responsible for severe infections in human and plant systems. They play significant roles in ensuring the balance of nutrients by recycling them in the atmosphere. Most of the discoveries related to genetics, molecular biology, physiology and biochemistry have been done by taking them as model systems. These are also used as host systems for industrial production of some compounds, which one may never associate with microbes; these are insulin, interferons, growth hormones, recombinant vaccines and many more. The interesting world of microbes is, therefore, worth exploring to the very depths – it is probable that they hold the key to human existence.

REFERENCES

Adler J. (1969). Chemoreceptors in bacteria. *Sci.* 166:1538-1597.

Avery OT, Macleod CM and McCarty M. (1944). Studies on the chemical nature of substance inducing transformation of Pneumococcal types. Induction of transformation by a deoxyribonucleic acid fraction isolated from Pneumococcus type III. *J. Exp. Med.*79:137-157.

Baltimore D. (1970). Viral RNA dependent DNA polymerase. *Nature.* 226:1209-1211.

Bamberg JH. (2002). British and global oil. 1950-1975. *The challenge of nationalism.* Vol.3 Cambridge Univ. Press. Pp.426-428.

Beadle G and Tatum E. (1941). Genetical control of biochemical reactions in *Neurospora crassa. Proc. Natl. Acad. Sci*. 27: 499-506.

Behring E and Kitasato S. (1890). Ueber das Zustandekommen der Diphtherie-Immunitat und der Tetanus-Immunitat bei thieren. [On the Establishment of Diphtheria Immunity and Tetanus Immunity in Animals.] *Deutsche medizinsche Wochenschrift.* 16:1113-1114.

Beijerinck M. (1899). Ueber ein contagium vivum fluidum als Ursache der Flechenkrankheit der Tabaksblatter. [About a contagium vivum fluidum as the cause of the lichen disease of the tobacco leaves.] *Centralblatt fur bacteriologie und Parasirenkunde*, Part II 5:27-33.

Beijerinck M. (1888). Die Bakterien de Papilionaceen knollchen. [The bacteria are the Papilionaceae nodules.] *Botanische Zeitung*. Vol. 46:725-804.

Brefeld O. (1872). *Botanissche untersuchungen uber schimmel pilze, hefti Mucor mucedo, chaetocladium jones ii, piptocephalis fresiana:zygomyceten.* [Mucor mucedo, chaetocladium jones ii, piptocephalis fresiana: zygomycetes, botanical studies on mold fungi] Leipzig.

Brock TD. (1967). Microorganisms adapted to high temperature. *Nature*. 214: 882-885.

Bruce D. (1895). *Preliminary report on the tsetse fly disease or Nagana in Zululand Durban,* Bennett & Davis.

Burril TJ. (1878). *Pear Blight. Trans. III.* State Hort. Soc. 114-116.

Chain E, Florey HW, Gardner AD, Heatley NG, Jennings MA, Orr-Ewing J and Sanders AG. (1945). Penicillin as a chemotherapeutic agent. *Lancet* 2:226-228.

Chamberland, C. (1884). Sur un filter donnant de l'eau physiologiquement pure. [On a filter giving physiologically pure water.] *Compt. Rend. Acad. D. Sc.* Paris xcix:247.

Charrin A and Roger J. (1889). Note sur le development des microbes pathogens dans le serum des animaux vaccines. [Note on the development of pathogenic microbes in the serum of vaccinated animals.] *Soc. De Biol.* 9e ser., I 667-669.

Cohn F. (1872). Ueber Bakterien, die kleinsten lebenden Wesen. [About bacteria, the smallest living beings.] *Luedritz'sche Verlagsbuchhandlung Carl Habe*, Berlin.

Cohn F. (1875). Untersuchungen ueber Bakterien. [Studies on bacteria.] *Beitrage zur Biologie der Planzen* 1;127-222.

Cramer A. (1870). *Dr. Muller Chem. Phy.* Beschreibung der Thermen von Baden in der Schweiz. [Description of the thermal baths of Baden in Switzerland.]

Currie JN. (1917). The citric acid fermentation of *Aspergillus niger*. *J. Bio. Chem.* 31:15-37.

D'Herelle F. (1917). Sur un microbe invisible antagoniste des *Bacillus dysentriques*. [On an invisible microbe antagonist of dysenteric bacilli.] *Comp. Rend. Acad. Sci.*165:373-375.

Darling SL. (1906). A protozoan general infection producing pseudotubercules in the lungs and focal necrosis in the liver, spleen and lymph nodes. *JAMA*. 46:1283-1285.

Dobell C. (1932). *Antonyvon Leeuenhoek and his animals*. London Staples Press. 1932. Republished. NY. Dover Publication Inc. 1960.

Dobell C. (1994). *Microbial process development*. World Scientific. pp205.

Domagk GJ. (1935). Ein Beitrag zur chemotherapie der bakteriellen infektionen. [A contribution to the chemotherapy of bacterial infections.] *Dtsch. med. Wochenschr.* 61:250-253.

Donovan C. (1903). On possibility of the occurrence of Trypanosomiasis in India. *Br. Med. J.* ii 79.

Duggar BM. (1948). Aureomycin; a product of the continuing search for new antibiotics. *Ann. NY Acad. Sci.* 30: 177-181.

Ehrlich P. (1881). Ueber das Methylenblau und seine Kliisch-bakterio skopische Verwerthung. [On the methylene blue and its clinical-bacterio scopic utilization.] *Ztschr. F. Klin. Med.* ii: 710-713.

Ehrlich P. (1912). Ueber laboratoriumsversuche und klinische Erprobung von Heinstoffen. [About laboratory experiments and clinical trials of ingredients.] *Chem. Ztg.* 36:637-638.

Elford, WJ. (1931). *A new series of graded colloidal membrane suitable for general bacteriological use especially in filterable virus studies*. 34: 505-521.

Escherich T. (1885). Die Darmbakterien des Neugeborenen und Sauglings. [The intestinal bacteria of the newborn and infants.] *Fortschr. d. Med.* 3:515-522.

Flemming A. (1929). On the antibacterial action of cultures of a Penicillium, with a special reference to their use in the isolation of *B. influenze*. *Brit. J. Exp. Path.* 10:226-236.

Gessard C. (1884). Classics in infectious diseases. On the blue green colouration that appears on bandages. *Rev.Infe. Dis.* 3:775-776.

Gram C. (1884). Ueber die isolirte Farbung der Schizomyceten in Schitt-und Trockenpreparaten [About the isolated staining of Schizomycetes in cut and dry preparations], *Fortschritte der medicin* 2:185-189.

Griffith F. (1928). The significance of Pneumococcal type. *J. Hyg.* 27:113-159.

Hasting JW and Gibson QH. (1967). The role of oxygen in the photoexcited luminescence of bacterial luciferase (*Photobacterium fischeri*) *J. Biol. Chem.*242:720-726.

Hayes W. (1952). Recombination in bacteria *Escherichia coli* K-12 unidirectional transfer of genetic material. *Nature* 169-118-119.

Hellriegel H and Wilfarth H. (1988). *Untersuchungen uber die Stickstoffnahrung der Gramineen und Leguminosen.* [Studies on the nitrogen nutrition of gramineae and legumes.] Beilageheft zu der Zeischrift des Vereins fur Rubenzucker-Industrie Deutschen Reichs. 234.

Hendriks J andBlume S. (2013). Measles Vaccination, Before the Measles-Mumps-Rubella Vaccine; *Am J Public Health.* August; 103:1393–1401.

Hensen GA. (1873). Undersogelser angaende spedalskhedens arsager [Investigations regarding leprosy's leprosy] (in Norwegian). *Norsk. Mag. Laegevid* 4:1-88.

Hensen GHA. (1973). The Man and his work. *Int. J. Leprosy,* 41:417-424.

Hershey A and Chase M. (1952). Independent function of viral protein and nucleic acid in growth of Bacteriophage. *J. Gen. Physiol.* 36:39-56.

Hiltner L. (1904). *Uber neuere Erfahrungen und problem aufdem Gebiete der Bodenbakteriologie under besonderer Berucksichtigung Arb.* [On recent experiences and problems in the field of soil bacteriology, with special consideration Arb.] DLG. 98:59-78.

Hitchins, Parker A and Leikind MC. (1939). The introduction of Agar-agar into bacteriology. *J. Bact.* 37:485-493.

Hungate RE. (1943). Experiments on cellulose digestion by protozoa in the rumen of cattle. *Bio. Bull.* 84:157-63.

Hungate RE. (1950). The anaerobic mesophilic cellulolytic bacteria. *Bact. Rev.* 14:1-49.

Hutchison CA III at J. Craig Venter Institute in La Jolla, Calif., and colleagues was titled, "Design and synthesis of a minimal bacterial genome."*Credit: C. Bickel /Science2016.*

Ishiwata S. (1905). Concerning sottokin a Bacillus of a disease of silkworm. *Rept. Asso. Seric.* Japan, pp.160-161.

Iwanovsky DI. (1892). On two diseases of tobacco. *Sel'.Khoz. lesov.* 169:108-121.

Jackson DA, Symons RH and Berg P. (1972). Biochemical methods for inserting new genetic information into DNA of simian virus 40: circular SV 40DNA molecules containing lamda phage genes and the galactose operon of *Escherichia coli. Proc. Natl. Acad. Sci.* 69:2904-2909.

Jacob F, Perrin D, Sanchez C and Monod J. (1960). Loperon Groupe de genes in a l'expression co or dome par un operateur. [Loperon Group of genes in expression by an operator.] *Compt. Rendu. Acad. Sci.* 245:1727-1729.

Ketchum PA. (1988). *Microbiology Concepts and Applications.* John Wiley and Sons. NY.

Kitasato S. (1889). Uber den tetnusbacillus. [About the tetnus bacillus.] *Ztschr. Hyg.u. Infektionskrank.* 7:225-234.

Kitasato S. (1894). The bacillus of bubonic plague. *Lancet* 2:428-430.

Kloepper JW and Schroth MN. (1978). Plant Growth promoting Rhizobacteria on radishes. *Proc. of 4th Int. conf. on Plant Patho. Bact. Angers.* France: Station de Pathologie Vegetale et Phytobacteriologie. INRA. 2:879-882.

Koch, R (1881). Zur Untersuchung von pathogenen Organismen. [For the investigation of pathogenic organisms.] *Mitth.a.d.Kaiseri.*

Koch, R. (1881). *Ueber Desinfection.* Mittheilungen aus dem kaiserlichen Gesundheitsamt [*About Disinfection.* Announcements from the Imperial Health Office] 1;234-282.

Koch R. (1884). *Die die isolirte Farbung der Schizomyceten in Schitt-und Trockenpreparaten fortschritte der Medicine* [*The isolated coloring of the Schizomycetes in Schitt and Trockenpreparaten advances the Medicine*] 2:1-88.

Koch R. (1882*). Die aetiologie der Tuberculose* [*The aetiology of tuberculosis*]. Berl. Klin. Wchnschr. Xix221-23.

Koning CJ. (1904.) Beitrage zur kenntnis des lebens dur humus pilze und der chemischen vorgange bei der humus bilding. [Contributions to the knowledge of the life of humus mushrooms and the chemical processes in humus education.] *Arch. Neerland Sci. Exact. Et nat.* 9:34-107.

Kornberg A. (1960). Biologic synthesis of deoxyribonucleic acid. *Sci.* 131:1503-1508.

Krasner RI. (1995) Pasteur: High Prist of Microbiology: Whereever he turned, Pasteur brought great insight to benefit huminity. *ASM News.* 61:575-578.

Lederberg J and Lederberg E. (1952). Replica plating and indirect selection of bacterial mutants. *J. Bacteriol.* 63:399-406.

Lederberg J and Tatum EL. (1946). Gene recombination in Eschericia coli. *Nature* 58:558.

Lederberg J, Cavallis LL andHayes W. (1952). Sex compatibility in *Escherichia coli. Genet.* 37:720-730.

Leishman WB. (1903). On the possibility of the occurrence of Trypanosomiasis in India. *Br. Med. J.* 1: 1252-1254.

Lemoigne M. (1926). Produits de deshydratation et de polymerization de Pacide b oxybutyrique. [Dehydration and polymerization products of bovine oxybutyric acid.] *Bul. Soc. Chim. Biol*. 8:770-782.

Leveran A. (1880). A new parasite found in the blood of malarial patients. Parasitic origin of malarial attack. *Bull. Mem. Soc. Med. Hos.* Paris 17:158-164.

Lewis T. (1878). The microscopic organisms found in the blood of man and animals and their relation to diseases. *Ann. Rpt. San. Commis. Govt. India Calcutta*. 14:157.

Luria S and Delbruck M. (1943). Mutation in bacteria from virus sensitivity to viral resistance. *Genetics* 28: 491-511.

Luria SE and Anderson TF. (1942). Identification and characterization of bacteriophages with the electron microscope. *Proc. Nat. Acad. Sci. US*. 28:127-130.

Lwoff A, Siminovitch L and Kjeldgaard N. (1950). *Comptes rendus* 231: 190-191.

Madigan MT, Martinko JM, Stahl DA and Clark DP. (2012). *Brock Biology of Microbiology 15th eds*. Pearson edu. Inc. San Fransisco.

Margalit J and Dean, D. (1985). The Story of *Bacillus thuringiensis* var. *israelensis* (*B.t.i.*). *Journal of the American Mosquito Control Association*, 1, 1–7.

Margulis L. (1968). Evolutionary criteria in Thallophytes: A radical alternative. *Sci.* 161:1020-1022.

McCrady MH. (1911). The numerical interpretation of fermentation tube results. *J. Inf. Dis.* 17:183.

Metchnikoff E. (1884). Ueber eine Sprosspilzkrankheit der Daphnien. Beitrag zur Lehre uber den Kampf der phagocyten gegen Krankheitserrenger, [About a budworm disease of the daphnia. Contribution to the teaching about the fight of the phagocytes against pathogens] *Archiv f. pathologische Anatomie und Physiologie und f. klinische Medicin*, 96:177-195.

Micheli PA. (1729). *Nova Plantarum*.

Montagnier LJ, Duquet LC, Axier C, Chamarat S, Neugeyre MT, Rey F, Barne Sinoussi F and Chermann JC. (1983) A new type of retrovirus isolated from the patients of presenting with lymphadenopathy and acquired immune deficiency syndrome: structural and antigenic relatednesswith equine infectious anemia virus. *Ann. Virol.* (Paris) 135:119-134.

Murray EGD, Webb RA and Swann MBR. (1926). A disease of rabbit characterized by a large mononuclear leucocytosis, caused by a hitheroto undescribed bacillus bacterium monocytogenes. *J. Pathol. Bacteriol.* 29:407-39.

Neisser A. (1879). Ueber eineder Gonorrhoe eigenthumliche Micrococcus form. [About one of the gonorrhea peculiar Micrococcus form.] Vorlanfige Mitteilung Cu. *F.D. Med. Wiss*. 28:497-500.

Nicolle, CH, Compte C and Conseil E. (1910). Experimental transmission of Exanthematous Typhus by body lice (*Pediculus vestimentii*). *Ann. Past. Inst*. 24:261-267.

Nobbe F. and Hiltner L. (1893). Impfet den Boden [Inoculate the soil] *Sachsische Landwirtsch Aftliche Zeitschrift.* 16:1-5.

Ochiai K., Yamanaka T., Kimura K. and Sawada O. (1959). *Nippion I Ji* 1961:34.

Ophuls W. and Moffitt H.C. (1900). A new pathogenic mold. *Philadelphia Med. J.* 5: 1471-1472.

Owen J.A., Punt J., Stranford S.A., Jones P.P. (2013). KUBY Immunology, Seventh Edition; Chapter 1; Overview of the Immune System (*A Historical Perspective of Immunology*) pp 2-9: W. H. Freeman and Company, NY.

Paradee A, Jacob F and Monod J. (1959). The genetic control and cytoplasmic expression of inducibility in the synthesis ofgalactosidase by *E. Coli. J. Mol. Biol.* 1:165.

Pasteur L. (1861). Animalcules infusoires vivant sans gaz oxygene II lre et determinant des fermentations. [Infusive animalcules living without oxygen gas II lre and determining fermentations.] *Compt. Rend. Acad. Sci.* (Paris) 52:344-347.

Pasteur L. (1880). De l'attenuation du virus cholera des poules. [Cholera virus attenuation of chickens.] *Compte. Rend. Acad. Se.* 91:673-680.

Peterson SN, Lucier T, Heitzman K, Smith EA, Bott KF, Hu PC, and Hutchison CA. (1995) 3rd Genetic map of the *Mycoplasma genitalium* chromosome; *J Bacteriol.* 177: 3199–3204.

Petri RJ. (1887). Eine kleine Modification des Koch'schen Plattenverfahrens. [A small modification of the Koch plate process.] *Centralbl. F. Bacteriol.* 1:279-280.

Pfeiffer R. (1894). Weitere Untersuchungen ueber das Chloraimmunitatund ueber specifisch bactericide Processe. [Further investigations on Chlora immunity and on specific bactericidal processes.] *Ztschr. F. hyg. U. Infektionskrankh* xviii:1-16.

Ricketts H. (1909). A microorganism which apparently has a specific relationship to Rocky Mountain Spotted fever. *J. Am. Med. Soc.* 52: 379-380.

Riley RL, Mills CC, Nyka W,Weinstock N, storey PB Sultan LU, Riley MC and Wells WF.(1959). Ariel dissemination of pulmonary tuberculosis. A two year study of contagion in a tuberculosis ward. *Amer. J. Hyg.* 70:185-196.

Rivers T. (1926). Filterable Viruses a critical Review. *J. Bact.* 14: 217.

Ross R. (1899). Mosquitoes and Malaria *Brit. Med. J.* 432-433.

Roux E and Yersin A. (1888). Contribution a l'etude la diphtheria. [Contribution to the study diphtheria.] *Ann. Inst. Pasteur* 2:273.

Ruska E and Knoll M. (1931). Diemagnetische Sammelspule fur scheelle Elektronenstrahlen. [the magnetic concentrating coil for fast electron beams]. *Z. techn. Physik* 12:389-400 and 448.

Safferman R.S. and Morris M.E. (1963) Algal virus: isolation. *Science* 140:679.

Schatz A, Bugie E and Waksman SA. (1944). Streptomycin, a substance exhibiting antibiotic activity against Gram positive and Gram negative bacteria. *Proc. Soc. Exp. Biol. Med.* 55:66-69.

Schaudinn FR & Hoffman E. (1908). Uber Spirochaetenbe funde in lymphobum sensaftsy phi litischer detsch. *Med. WS chr.* 31: 711.

Schenk BR. (1898). On refractory subcutaneous abscesses caused by a fungus possibly related to the sporotricha. *John Hopkins Hos. Bull.* 9:286-290.

Schloesing J and Muntz A. (1877). Sur la Nitrification par les ferments Organises. [On Nitrification by organized enzymes.] *Comptes Rendus del'Academie des Science*, Paris, LXXXIV:301-303.

Schwarz HP, Dorner F. (2003). Landsteiner K and his major contributions to haematology. *British J. of haematology;* 121:556–565.

Smith DE and Ames BN. (1966). Phosphoribosyladenosine monophosphate, an intermediate in histdine biosynthesis. *J. Biol. Chem.* 240:3056-3063.

Smith HO and Wilcox KH. (1970). A restriction enzyme from *Haemophilus inflenzae*: I Purification and general properties. *J. Mol. Biol.*51: 379-379.

Smith T & Kilbourne FL (1893) Investigations into the nature, causation and prevention of Texas or southern cattle fever. *Bur. Anim. Bull.* 1:151-152.

Smith T and Kilbourne L. (1893). Investigations into the nature causation and prevention of Texas or southern cattle fever. *Bur. Anim. Ind.Bull.* 1:151-152.

Sohngen NL. (1906). Ueber Bakterien welche Methan als Kohlenstoffnahrung and energiequelle gebrauchen [About bacteria that use methane as carbon food and energy source]. *Zentralbl. Bakteriol. Parasitik.* Abt. I. 15:513-517.

Stainer Y, Ingraham JL, Wheelis ML and Painter PR. (1987). *General Microbiology Vth Ed.* Macmillan Press Ltd. London.

Staley JT, Gunsalus RP, Lory S and Perry JJ. (2007). *Microbial life.* 2nd ed.Sinauer Associates Inc. Sunderlaqnd Massachsetts.

Stanley W. (1935). Isolation of a crystalline protein possessing the properties of tobacco-mosaic virus. *Sci.* 81:644-654.

Temin HM and Mizutani S. (1970). RNA dependent DNA polymerase in virions of Rous sarcoma virus. *Nature* 226:1211-1213.

Tildane JE. (1935). The Algae and their life relations.Jackson DA, Symons RH and Berg P. (1972). Biochemical method for inserting new genetic information into Simian virus 40: DNA molecule containing lambda phage gene and galactose operon of *E. coli. Proc. Natl. Acad. Sci.* 69:2904-2909.

Tillet M. (1755). *Dissertation sur la cause qui corrompt et noircit les grain dela cause deble dan les epis et surles moyens de prevenir Ces accidents* [*Dissertation on the cause which corrupts and blackens the grain of the cause in the ears and on the means of preventing these accidents*].

Twort F. (1915). An investigation on the nature of the ultra microscopic virus. *Lancet.* 2; 1241-1243.

Tyndall J. (1877). On heat as a Germicide when discontinuously applied. *Proc. Royal Soc.* London 25:569.

Ugalde UO and Castrillo JI (2002). *Applied Mycology and Biotechnology*. Vol. 2: Agriculture and Food production.

Van Neil CB. (1931). On Morphology & physiology of the Purple and Green Sulphur bacteria. *Arch. Mikrobiol* 3: 1-112.

Wakesmann, SA. (1954). *My Life with the Microbes*. Simon and Schuster. NY. Pp. 145.

Walter and Angelina. (1882). Early contributors to Bacteriology, *ASM News*. 581992p425 Link to ASM news Article.

Walter Reed. (1911). Papers by various authors of the yellow fever commission. *Sen. Doc*. 822. Washington, D.C.

Warren JR and MarshallB. (1984). Unidentified curved Bacilli on gastric epithelium in active chronic gastritis. *Lancet* 1:1273.

Welch WH & Nuttall G. (1892). A gas producing bacillus (Bacillus aerogenes capsulatus, nov. spec.) capable of rapid development in the blood vessels after death. *John Hopkins Hosp. Bull*. 3:81-91.

Winogradsky, S. (1887). Uber Schwefelbacterien. [About sulfur bacteria] *Botanische Zeitung* XLV:489-507.

Winogradsky, S. (1890). Recherches sur les Organismes de la Nitrification. [Research on Nitrification Organisms.] *Computer Rendu* Vol. 1101013-1016.

Woese CR and Fox GE. (1977). Phylogenetic structure of prokaryotic domain. The primary kingdom. *Proc. Natl. Acad. Sci*. 74:5088-5090.

Wollman E and Jacob F. (1957). Sur les processus de conjugaison et de recombinaison chez [On the processes of conjugation and recombination in] *Escherichia coli*. *Ann. Inst. Pasteur* 93:323-339.

Wood HG and Werkman CH. (1936). The utilization of CO_2 in the dissimilation of glycerol by Propionic acid bacteria. *Biochem J*. 30: 48-53.

Woodruff, A and Good Pasteur E. (1931). The susceptibility of chorioallantoic membrane of chick embryo to infection with the fowl pox virus. *Am. J. Path*. 7:209-222.

Woods DD. (1940). The relation of p-aminobenzoic acid to the mechanism of the action of sulfanilamide. *Brit. J. Exp. Path*. 21:74-90.

Yanofsky C, Carlton BC, Guest JR, Helinsky DR and Henning U. (1964). On colinearity of gene structure and protein structure. *Proc. Natl. Acad. Sci*. 51:266-274.

Yersin A. (1894). La peste bubonique a Hong Kong. *Ann. Inst. Pasteur* 8: 662-667.

In: Recent Advances in Microbiology. Volume 3 ISBN: 978-1-53614-057-6
Editors: N.C. Gautam, S.P. Tiwari and R. Sharma

Chapter 2

MICROBES AND THEIR METABOLITES AS A RICH SOURCE OF ANTIOXIDANTS: AN OVERVIEW

Meenakshi Tiwari[1,*] and Om Prakash Narayan[1,2]
[1]Department of Biochemistry,
V. B. S Purvanchal University Jaunpur, India
[2]Plant Biotechnology, Rajiv Gandhi South Campus,
Banaras Hindu University, Barkachha, Mirzapur, India

INTRODUCTION

Oxygen is an important gaseous component of the environment and is extremely essential for the living systems. Beyond the physiological levels it may show deleterious effects in the biological system. When oxygen gets converted to partially reduced forms, it results in production of Reactive oxygen intermediates (ROI). Along with this reactive nitrogen species (highly unstable molecules of nitrogen and oxygen), are capable of performing unrestricted oxidation of several cellular macromolecules and producing oxidative damage of cell. These oxidants are either produced exogenously or endogenously (Figure 1). The exogenous sources include exposure to various oxidants in atmosphere like diesel exhaust, cigarette smoke, ozone, nitrogen dioxide, sulfur dioxide, drugs and radiations.

*E-mail: meetmt@gmail.com.

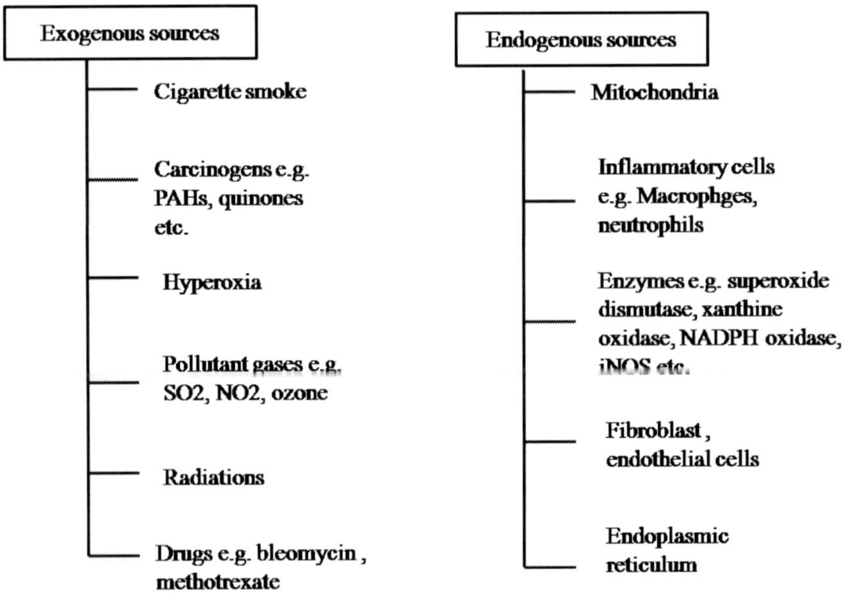

Figure 1. Exogenous and endogenous sources of antioxidants.

Along with exogenous sources there are several endogenous sources which give rise to various oxidant species inside the biological system. There is production of ROIs during normal metabolism processes. Sources include mitochondria, microsomes and several enzymes like xanthine oxidase, P450 monooxygenase, lipooxygenase, cyclooxygenase, indole amine dioxygenase, monoamine oxygenase etc.

Various inflammatory cells like macrophages, monocytes, eosinophils and neutrophils also generate microbicidal or pathological levels of superoxide ($O_2^{\cdot-}$) through the action of enzyme NADPH oxidase, for microbicidal and other cellular defence mechanism against foreign particles. The superoxide generated is dismutated to hydrogen peroxide (H_2O_2) by the action of superoxide dismutase enzyme (Kirkham & Rahman, 2006). This H_2O_2 is also produced by the action of various other enzymes like monoamine, amino acid oxidase and xanthine oxidase.

Peroxidase enzymes like eosinophil peroxidase (EPO) and myeloperoxidase (MPO) are important in host defense against parasites and pathogens, as they amplify the response of H_2O_2 by forming deleterious HOCl and HOBr (Henricks & Nijkamp, 2001). NO is another important free radical produced by the action of enzyme nitric oxide synthase on L-arginine resulting in the formation of L- citrulline and NO. This NO further reacts with superoxide ion to produce more deleterious peroxynitrite radical which causes damage to tissues. In plants exposure to drought stress, salt stress, chilling, heat shock, heavy metals, mechanical stress, nutrient deprivation, pathogen attack and high light stress also result in ROI generation via photorespiration, mitochondrial respiration, glycolate oxidase pathway in peroxisomes and NADPH oxidase system.

To combat with these reactive species body contains a strong defense mechanism consisting of several antioxidants. Antioxidants are defined as the substance, that even present in low concentrations (compared to those of an oxidisable substrate), significantly delays or prevents oxidation of that substance (Halliwell et al., 1995; Huang et al., 2005). The concept of antioxidants *in vivo*, includes antioxidant enzymes, iron binding and transport proteins, other compounds affecting signal transduction and gene expression (Gutteridge, 1989). These are also of two types:

a) Non enzymatic antioxidants- These consists of urate, reduced GSH, vitamin E and C, extracellular SOD, catalase, GSH-Px, GSH, mucopolypeptideglycoproteins, cerruloplasmin, Fe binding proteins e.g., lactoferrin and transferrin and bilirubin (Avissar et al., 1996; Comhair & Erzurum, 2002; Rahman et al., 2006).
b) Enzymatic antioxidants- Along with the non enzymatic antioxidants several enzymatic antioxidants are also required which either act alone or synergistically against oxidative stress. The major enzymatic antioxidants present in lungs are superoxide dismutases, catalase, glutathione peroxidase and small molecular weight redox proteins (thioredoxins, peroxiredoxins and glutaredoxins, heme oxygenase-1) which play a crucial role in the lung antioxidant defence mechanisms.

The antioxidants present in cells, cell membranes and extracellular fluids are up-regulated and mobilized to neutralize the excessive ROS formed during oxidative insults (Lygoe et al., 2007). Any deficiency in the antioxidant defense of the body, to combat with these oxidants produced, leads to oxidative damage to various macromolecules (protein, lipids, DNA) and this condition of imbalance is known as oxidative stress (Mikulikova & Popes, 2001). Oxidative stress has been found to be involved in the pathophysiology of various diseases including neurodegenerative diseases like Alzheimer's disease, Parkinson's disease, diabetes, rheumatoid arthritis, cardiovascular diseases, lung inflammatory disorders like asthma acute respiratory distress syndrome, chronic obstructive pulmonary disease and neurodegeneration in motor neuron diseases. Antioxidants acts to reduce/ inhibit oxidants by various mechanisms like:

a) They inhibit the formation of free radicals in the initiation step of auto-oxidation
b) Interrupt the propagation of the free-radical chain.
c) They can chelate free metal ions in solution
d) Act as reducing agents (electron donors) and remove or reduce concentrations of oxygen.

For many decades, there has been a growing interest in finding novel antioxidants. Natural antioxidants isolated from plants, fungi and marine algae represent the most useful nutraceuticals and functional foods for health protection and disease prevention (Gutteridge & Halliwell, 1994). In the search of new natural antioxidants, a number of bioactive compounds obtained from plants, algae and microorganisms have been demonstrated to possess potent antioxidant activity. Further, there is an overview of some microbes or their secondary metabolites which are studied for their antioxidant potential.

Bacteria or Their Extracts as Antioxidants

Bacteria are a large group of prokaryotic microorganisms which are present in almost every habitat including soil, digestive tract of animals, hot springs, water, organic matter etc. These microorganisms are also emerging as a source of natural antioxidants. Several marine animals are reported to have high polyunsaturated content, indicating the presence of certain antioxidants in their body, which prevents their oxidation. A study by Takao et al., 1994 reported 112 bacterial isolates from sterilized sea water homogenates of gills, viscera and scales of fish and shellfish, which were showing positive antioxidant screening tests with DPPH. The effective antioxidants compounds which can be added to food components are not in vast number. Food industry is using several synthetic antioxidants, such as butylated hydroxyanisole (BHA), butylated hydroxytoluene (BHT), and tertiary butylhydroquinone (TBHQ), as food additives to prevent lipid peroxidation. Due to toxic and carcinogenic components of health concern formed during the degradation of these antioxidants their use is limiting (Maeura et al, 1984; Ito et al, 1985).

So, the need for exploration of safer, more effective and economic natural antioxidants is of great advantage. Ongoing researches are also targeting on exploring certain antioxidants of microbial flora origin, which can grow on fatty medium and protect the fat from oxidation (Lea, 1938). The oil soluble fraction of *Mycobacterium phlei* showed antioxidant activity when it was added to cotton seed oil (Shappirio, 1944). A study by Smith and Alford, 1970, showed that a considerable amount of antioxidant activity was observed by methanol soluble-benzene soluble fractions of *Bacillus cereus, Lactobacillus dextranicum, Micrococcus freudenreichii* and *Sarcina lutea*, while methanol fractions of three *Pseudomonas* species showed considerable activity that could not be extracted with benzene. In this study, increase in the period required for the initiation of rancidity (indicated by changes in peroxide value) was taken as a measure of antioxidant activity. Two intestinal lactic acid bacteria *B. longum* and *L. acidophilus* in intact cellular form and intracellular cell-free extracts, demonstrated antioxidative activity. This was demonstrated by inhibition of linoleic acid peroxidation and scavenging activity of DPPH free radical (Lin & Chang, 2000). In another study, antioxidative activity of *L. casei* ssp *casei* was reported *in vitro* by linoleic acid peroxidation assay and MS-TBAR assay. Also the strain was able to lower cholesterol during *in vivo* study and thus can be used as a good dietary supplement (Kapila et al., 2006).

Sand biological soil crusts are potential target of damage due to oxidative stress caused by drought and irradiation. The bacteria residing in these extreme conditions are found to contain significant amount of antioxidant substances to combat with oxidative stress. *Bacillus simplex* XJ-25, isolated from sand biological soil crusts in Gurban Tonggut Desert, Xinjiang, China has also been reported to have radical scavenging activity (Wang et al., 2011). Exopolysaccharide extracted from *Enterococcus faecium* isolated from Ngari, a traditional fermented fish of Manipur have showed strong DPPH and superoxide radical scavenging activity (Abdhul et al., 2014). A study was conducted to see the protective effect of strains of *Lactobacillus* and *Bifidobacterium* on gingival fibroblasts. *Bifidobacterium breve* A28 showed the highest Exopolysaccharide production (122 mg/l) and remarkable antioxidant activity, which were demonstrated by its ability to scavenge α,α-diphenyl-1-picrylhydrazyl free radical, inhibit lipid peroxidation and chelate iron ion. This suggested that LAB with antioxidative activity could be a good natural therapy agent for periodontal disorders (Mendi et al., 2014). The bioindigo pigment produced by *Pseudomonas* sp. HAV-1 have shown ascorbic acid equivalent antioxidant property (Dua et al., 2014). In a study, Lactic acid bacteria *Lactobacillus plantarum* KCC-24 (L. plantarum KCC-24), was isolated and characterized from Italian ryegrass (*Lolium multiflorum*) forage which showed potent antioxidant, antifungal and hydrogen peroxide resistant property (Vijayakumar et al., 2015). A study was performed, in which *Lactobacillus gasseri* SBT2055 (LG2055) was feeded to *C. elegans*. This was found effective to extend the lifespan in *C. elegans* by strengthening the resistance to oxidative stress and by stimulating the innate immune response signaling including p38MAPK signaling pathway and others (Nakagawa et al., 2016). The antioxidant activities in vitro of *Lactobacillus plantarum* MA2, isolated from Chinese traditional Tibetan kefir grains was evaluated. The results showed that *L. plantarum* MA2 fermentate (fermented supernatant, intact cell and cell-free extract) had strong reducing capacities, lipid peroxidation inhibition capacities, Fe^{2+} chelating abilities, as well as various free radical scavenging capacities (Tang et al., 2017). Hetero-exopolysaccharide secreted by *Bacillus tequilensis* FR9 (BtqEPS), exhibited antioxidant effects by scavenging reactive oxygen species (ROS) and demonstrated strong reducing power (Rani et al., 2017). Findings of a recent study demonstrated that biosurfactants from *L. casei* strains exhibited considerable antioxidant potencies as indicated by DPPH scavanging activity (Merghni et al., 2017). In a study by Jan et al., the compounds, sarcovagine-D, alkaloid-C, and holaphylline, isolated from *Sarcococca saligna* were evaluated for their *in vivo* hepatoprotective potential against CCl_4. The decrease in the activity of hepatic antioxidant enzyme (SOD) was significantly prevented by holaphylline, likewise gradually the levels of MDA and GSH were also normalized compared to silymarin (Jan et al., 2017). The strain *Streptomyces coelicoflavus* BC 01 was isolated in a study from mangrove soil and used as inoculum for submerged fermentation. This was purified and three pure bioactive compounds named as BC 01_C1, BC 01_C2 and BC 01_C3 were isolated. These compounds also reported to possess *in- vitro* antioxidant and anti-inflammatory activities (Raghava et al., 2017).

FUNGI AND ITS METABOLITES AS ANTIOXIDANTS

Fungi is a very diverse group of microorganisms which consists of around 1.5 million species. These species provide various useful metabolites like flavanoids, phenols, alkaloids, tetralones, benzoquinones, steroids, terpenoids, and xanthones. The exploitation of these bioactive compounds make them potential therapeutic target in pharmaceutical industry and food industry (Archer, 2010). Recent researches have highlighted various fungi or their secondary metabolites as a good source of antioxidants (Rodrigues et al., 2005; Arora & Chandra, 2010). Soil isolate of *Aspergillus fumigates* have been reported to possess great antioxidant activity (Arora & Chandra, 2011). A study by Masalu et al., 2012 on free radical scavenging activity of crude extract of indogenous fungi of Tanzania, have also highlighted certain species of marine ascomycetes, domestic zygomycetes and forest basidiomycetes as potential antioxidant sources. This can supplement human diet after further research. *Cordyceps sinensis* (Berk.) Sac, commonly known as Dong-Chong-Xia-Cao (winter worm-summer grass) in Chinese, is an important herbal medicine used in china for various ailments. Hot water extract of *Tolypocladium* sp. of anamorphic fungus was isolated from wild *C. sinensis* and was reported for its antioxidant activity by Zheng et al., 2008. More detailed research on *Tolypocladium sp.* can increase the use of antioxidant property of this fungus over the endangered *C. sinensis*. A study by Bhanja et al., 2009 have showed that fermented wheat grain have more antioxidant potential. Study demonstrated that wheat grain fermented with *Aspergillus oryzae* and *Aspergillus awamorinakazawa* showed enhanced phenolic content and antioxidant property.

Yellow fungal dye produced by a strain of *Epicoccum nigrum* has shown antioxidant property. It also have exhibited property to colour the food emulsion and food product without using synthetic dyes. Thus limiting the eventual harmful effects of synthetic dye and increasing consumer interest (Cretu et al., 2011). Secondary metabolites collected in ethanol extract of *Picnoporus sanguineus* fungus, which were cultivated on potato dextrose broth and peptone medium, also demonstrated antioxidant property (Borderes et al., 2011).

Another important type of microorganisms, endophytic fungi which are present hidden in healthy host plants, are also an abundant source of new beneficial bioactive compounds. These active principles of endophytic fungi, have presented themselves as potential targets for exploitation in pharmaceutical, agricultural and industrial area (Tan & Zou, 2001). Along with this researches have reported their antiviral, anticancer and antimicrobial effects (Guo, 2000; Strobel et al., 1996). Huang et al., 2007 studied 292 morphologically distinct endophytic fungi cultures isolated from 29 traditional Chinese medicinal plants for total phenolic content and total antioxidant activity. Antioxidant potential of these enophytic fungi is also reported in various studies some of which are present in the table above (Table 1a).

Table 1a. Some endophytic fungi with antioxidant property

Endophytic fungi	Host plant	Antioxidant property	Reference
Phyllosticta sp.	*Guazuma tomentosa* H. B and K (Sterculiaceae)	Antioxidant activity against DPPH and ABTS radicals	Srinivasan et al., 2010
Alterneria alternate	*Coffea Arabica L.*	DPPH radical scavenging	Fernandes et al., 2009
Berkleasmium sp. Dzf12 (polysaccharides)	*Rhizomes of Dioscorea zingiberensis,*	DPPH, Iron chelating, hydroxyl radical scavenging	Li et al., 2011
Phomopsis sp. and Xylaria sp	*Emblica officinalis*	Reducing power and DPPH scavenging	Nath et al., 2012
Phomopsis sp. GJJM07	*Mesua ferrea*	DPPH scavenging	Jayanthi et al., 2011
7 genera isolated (Hypocrea, Penicillium, Tolypocladium, Chaetomium, Xylaria, Nemania, and Creosphaeria)	*Scapania verrucosa*	TEAC, DPPH, ABTS scavenging activity	Zeng et al., 2011
Phomopsis sp, Diaporthe phaseolorum, Phomopsis sp., Diaporthe sp, Sordariomycetes sp.	*Costus spiralis*	PPH, FRAP scavanging activity	Marson et al., 2014
Aspergillus sulphureus	*Sida acuta*	DPPH free radical scavenging activity, anticancer and DNA fragmentation against EAC cells	Murali et al., 2017

An antioxidant polysaccharide was isolated from the broth of marine filamentous fungus *Keissleriella* sp. YS 4108. The compound was found protective against superoxide radical, DNA damage, Copper induced LDL peroxidation and possessed metal chelaing property. Thus, it is a potential source of antioxidant which can have therapeutic value against many diseases where oxidative stress plays an important role (Sun et al., 2004). Two major antioxidative compounds, variegatic acid (3,3′,4,4′-tetrahydroxy puluvinic acid) and an orange-red pigment, were isolated from *Suillus bovinus (L: Fr.) O. Kuntze*, and were reported for their antioxidative activity (Kasuga et

Table 2. (Continued)

Mushroom or its isolated product	Antioxidant property	Reference
Morchella spp.	Protection against ethanol induced gastrin mucosal lesions due to increased antioxidant response	Wei et al., 2011
Agaricus essettei, Agaricus bitorquis and Agaricus bisporus	In vitro antioxidant, anticholinesterase and antimicrobial activities	Öztürk et al., 2011
Lentinus lepideus	Antioxidant and antityrosinase activity	Yoon et al., 2011
Hericium erinaceus oligosaccharide	DPPH and ABTS radical scavenging activity	Hou et al., 2015
Polyporus sulphureus, Macrolepiota procera, Lycoperdon perlatum and Gomphus clavatus	Antioxidant activity and total phenolic and flavanoid content	Sarikurkcu et al., 2015
Flammulina velutipes polysaccharide I-A	Scavenge hydroxyl radical, superoxide anion and possessed reducing power with immunomodulatory potential	Wu et al., 2014
Sulfate substitution (DS) derived from Ganoderma atrum.	Immunomodulator and free radical inhibitor	Chen et al., 2015
Agaricus brasiliensis and Ganoderma lucidum	Increased activity of antioxidant enzymes and reduction of TBARS level in diabetic rats	Yurkiv et al., 2015
Agaricus bisporus polysaccharides (ABP)	Scavenging of DPPH radicals, hydroxyl radicals, ferric reducing activity	Yin et al., 2015
Exopolysaccharide (EPS) from Climacodon septentrionalis	Antioxidant property	Jia et al., 2015
Hericium erinaceus (HE)	Modulation of MMP-9/NF-κB and Nrf2-antioxidant signaling, antiinflammatory and anticancerous activity	Chang et al., 2016
Phellinus rimosus	Protection against acute alcohol-induced lipid peroxidation in the liver, kidney, and brain	Ajith et al., 2015
5 Tunisian mushrooms-Phellinus torulosus, Fomes fomentarius, Trametes versicolor, Pisolithus albus, and Fomitopsis pinicola	Total phenolic content and antioxidantactivity per the DPPH and metal-chelating activity assays	Khadhri et al., 2017
Lentinus squarrosulus, L. tuber-regium, and Macrocybe gigantean	Free radical scavenging, reducing power, and chelating effect	Das et al., 2017
Cordyceps (C. gracilis, C. cicadae, C. sinclairii) and Metacordyceps (M. dhauladharensis)	Antioxidant and antibacterial activity	Sharma & Gautam, 2017
Boletus edulis	Antioxidant (2,2-diphenyl-l-picrylhydrazyl [DPPH], nitric oxide, super-oxide anion*, and hydroxyl radicals and ferric-reducing antioxidant power) and antiproliferative MTT assays (human breast MCF-7 cancer cell line)	Novakovic et al., 2017
Birch polypore Piptoporus betulinus	Antioxidant, antimicrobial, cytotoxic, and angiotensin I-converting enzyme (ACE) inhibitory activity	Vunduk et al., 2015
Ganoderma lucidum (MTCC1039)	Antioxidant activity and total phenolic content	Rani et al., 2015

Mushroom or its isolated product	Antioxidant property	Reference
Water-soluble polysaccharides (ORWP) and alkali-soluble polysaccharides (ORAP), prepared from the mushroom Oudemansiella radicata	DPPH radical scavenging activity, hydrogen peroxide scavenging activity and lipid peroxidation inhibitory effects	Liu et al., 2017
Exopolysaccharides (ExPSs) and endopolysaccharides (EnPSs) were isolated from mycelia and culture medium of Agaricus brasiliensis	Antioxidant and antiaeging activity	Zhang et al., 2016
Pleurotus giganteus (Berk. Karunarathna and K.D. Hyde)	Suppression of the STAT 3 and COX-2 pathways in LPS stimulated RAW 264.7 cells and its potential as endogenous antioxidant stimulant.	Baskaran et al., 2017
Agaricus urinascens (Jul. Schäff. & F.H. Møller) Singer and Volvopluteus gloiocephala (DC.) Vizzini, Contu & Justo	Antioxidant activity	Ekin et al., 2015
Ophiocordyceps sinensis CS1197	Antibacterial and antioxidant activity	Shashidhar & Manohar, 2016

Algae, Blue Green Algae and Their Metabolites as Antioxidants

Cyanobacteria are gram negative, photosynthetic prokaryotes and important contributors to the nitrogen economy of soil. Not with standing with above, cyanobacteria has developed antioxidative defence system to detoxify ROS and protects from oxidative stress. *Spirulina* is a blue green algae, rich source of protein and multiple vitamins (Fahmy et al., 2017). These organisms also produce natural antioxidant such as carotenoids and xanthophylls. Furthermore, *Phormidium tenue* (KMD 33) showed antioxidant property due to presence of phycocyanin and carotenoid (Naga Satya & Thajuddin, 2008). The α-TOC is a non-enzymatic antioxidant of cyanobacteria and is identified to check lipid peroxidation, chlorophyll photooxidation (Simontacchi et al., 1993), thymine dimerization and formation of singlet oxygen (McVean & Liebler, 1999). Moreover, deciphering the molecular basis of antioxidant chattels of cyanobacteria, functional genomics and proteomics has been used to identified genes and their respective proteins (Narayan et al., 2016). Under iron starvation-induced production of reactive oxygen species in *Anabaena* sp. PCC 7120, Fe/Mn SOD, catalase and peroxiredoxin (Prx) was up regulated, which helped in survival of organisms (Narayan et al., 2011). Up-regulation of catalase (CAT), peroxiredoxin (Prx), thioredoxin (Trx) and oxidoreductase appears to protect the cells from oxidative stress (Pandey et al, 2012). Recently Narayan et al., 2010 reported that *all 3940* gene of *Anabaena* sp PCC 7120, a DNA binding protein (Dps) scavenge the ROS in *E. coli* under nutrient starvation. This is also reported to protect the *Synechococcus* cells from oxidative stress to *Thermosynechococcus elongatus* (Franceschini et al, 2006; Peña et al, 1995). Similarly, *E. coli* transformed with *pcs* (Chaurasia et al, 2008) and *ahpc* (Mishra et al, 2009) genes from *Anabaena* sp PCC 7120 offer protection against oxidative stress. Irrigation of wheat crop with sea water creates salt stress. A study by Haana et al., 2008 (Haana et al., 2008) have reported that applying *Chlorella ellipoida* and *Spirulina maxima* extracts increased the antioxidant defense and stress tolerance of the wheat crop when irrigated with sea water.

Table 3. Some algae and its extracts with antioxidant properties

Algae or its extracts	Antioxidant property	Reference
Irish seaweeds; Laminaria digitata, Laminaria saccharina, Himanthalia elongata, Palmaria palmata, Chondrus crispus and Enteromorpha spirulina	DPPH scavenging, total phenolic content, total flavanoid	Ahmadi et al., 2015
Methanol, dichloromethane and hexane extracts, as well as the essential oils of brown algae (Phaeophyta) Colpomenia sinuosa, Dictyotadichotoma, Dictyota dichotoma var. implexa, Petalonia fascia and Scytosiphonlomentaria.	ABTS radical scavenging, β carotene bleaching assay, total phenolic	Tosin et al., 2017
Red alga, dulse (Palmaria palmata) and three kelp (Laminaria setchellii, Macrocystis integrifolia, Nereocystis leutkeana) extracts	Reducing activity using ascorbic acid and H_2O_2, total polyphenols	Romay et al., 1998
Water and ethanolic extract of Sargassum boveanum	DPPH radical scavenging activity, total phenolics	Xie et al., 2016
Synechococcus sp. FACHB283, Chlamydomonas nivalis and Nostoc ellipsosporum CCAP 1453/17, Chlorella sp.	Trolox equivalent antioxidant activity	Benedetti et al., 2004
Phaeodactylum tricornutum, Nannochloropsis sp., Isochrysis sp.,Tetraselmis suecica, Chaetoceros calcitrans, Schizochytrium sp.,Chlorella sp. 1 and sp. 2 Porphyridium cruentum, Botryococcus braunii, Tetraselmis sp., Nannochloropsisoculata, Isochryis sp. and Neochloris oleoabundans (Chlorophyta)	TEAC, FRAP assay, carotenoid content, phenolic content	Yen et al., 2001
Spirulina maxima	In vivo antioxidant activity in liver and plasma (inhibition of lipid peroxidation), In vitro tests in brain homogenate, phenolic, β carotene, α tocopherol content measurement	Cox et al., 2010
Methanol and aqueous extract Chlorophyta Ulva lactuca (U), Phaeophyta Sargassum crassifolia (S) and Rhodophyta Digenea simplex	Lipoxygenase, DPPH, Ames test	Demirel et al., 2010
Scytosiphon lomentaria, *Papenfussiella kuromo*, *Nemacystus decipiens* and one Rhodophyceae *Porphyra* sp.	Reduction of hemoglobin-induced linoleic acid peroxidation, reducing power, ferrous ion chelating, 1,1-diphenyl-2-picrylhydrazyl (DPPH) radical-scavenging and scavenging of a superoxide anion activity	Yuan et al., 2006

A wide variety of applications have been reported now a days of algal biomass as well as algae-derived compounds including its role as animal feed, aquaculture, source of food and nutrition and as potential health promoting components of diet (Chkhikvishvili & Ramazanov, 2000; Huang & Wang, 2004). During the process of food production, process and storage the endogenous antioxidants are lost in some extent resulting in less protection to lipid peroxidation. Also due to the great health benefits of antioxidants of natural origin on human health the demand of antioxidants in food industry is also increasing day by day. Algae and microalgae is emerging as an important natural source of antioxidants that can be used in food industry for functional food production (Michalak et al., 2017). These organisms are consumed as food component in various parts of world. Also they are reported to show therapeutic potential against various ailments (Borowitzka, 1999; Mendes et al., 2003; Cohen, 1999; Grima et al., 2003; Ahmadi et al., 2015; Tosin et al., 2017). C-Phycocyanin a protein bound pigment found in blue green algae has been reported for its antioxidant and anti-inflammatory activity (Romay et al., 1998; Xie et al., 2016). Phycocyanin extracted from *Aphenizomenon flos-aquae*, a fresh water blue green alga have also been reported for antioxidant activity (Benedetti et al., 2004). Three compounds 3,3''-dihydroxyterphenyllin, 3-hydroxyterphenyllin, and candidus in B were isolated from broth filtrate of *Aspergillus candidus* (CCRC 31543) and reported to have antioxidant property (Yen et al., 2001). Some more examples of algae showing antioxidant property is given in the table below (Table 3).

During the production of algae essence from microalgae *Chlorella vulgaris*, algae protein waste is produced as by-product which is used as animal feed. Zahra et al., in 2009 hydrolyzed this algae protein waste using pepsin and isolated a peptide VECYGPNRPQF which showed quenching capacity of free radicals, like hydroxyl radical, superoxide radical, peroxyl radical, DPPH radical and ABTS radicals. Thus it can be of great benefit as a potent antioxidant.

Plants and their active principles have been used since long back for cure of various ailments. Various species and their bioactive compounds have been constantly, scientifically validated for their strong antioxidant potential. But overharvesting of their non fruit part becomes environmental concern. So, microbes are the next most potential source for antioxidant which are present as numerous species which are still to be explored for their antioxidant and therapeutic potential. Thus, microbes or their secondary metabolites are emerging as a potential source of natural antioxidants. There natural origin make them safe as compared to the synthetic antioxidants BHT and BHA, which are reported to produce carcinogenic metabolites on degradation. Also, the heterotrophic nature of microbes make them easier to grow on various substrates. The production of microbes or their metabolites can be easily scale up at industrial level production. Edible nature and antioxidant potential of some microbes make them of great use in food industry as additives or as supplements to delay or reduce the oxidative damage during storage and processing. Hence, microbes are of great concern for research with lots of nutraceutical, antioxidant and therapeutic properties to be explored.

Acknowledgments

Dr. Meenakshi Tiwari and Dr. Om Prakash Narayan are thankful to Dr. S. P. Tiwari and Dr. R. Sharma for critical observation and suggestion during preparation of manuscript. We are also thankful to Head of Department for providing facility.

References

Abdhul K, Ganesh M, Shanmughapriya S, Kanagavel M, Anbarasu K, Natarajaseenivasan K. Antioxidant activity of exopolysaccharide from probiotic strain *Enterococcus faecium* (BDU7) from Ngari. *Intl J Biol Macromol.* 2014;70:450-4.

Ahmadi A. Zorofchian MS, Abubakar, S, Zandi K. Antiviral Potential of Algae Polysaccharides Isolated from Marine Sources: *A Review. Biomed. Res. Int.* 2015; 825203.

Ajith TA, Janardhanan KK. Medicinal Mushroom Cracked-Cap Polypore, *Phellinus rimosus* (Higher Basidiomycetes) Attenuates Acute Ethanol-Induced Lipid Peroxidation in Mice. *Int J Med Mushrooms.* 2015;17(11):1061-7.

Archer DB. Filamentous fungi as microbial cell factories for food use. *Curr Opinion Biotechnol* 2010; 11(5): 478–483.

Arora DS, Chandra P. Antioxidant activity of fungi isolated from soil. *Brazilian J Microbiol* 2010; 41: 465–477.

Arora DS, Chandra P. Antioxidant Activity of *Aspergillus fumigatus. ISRN Pharmacol* 2011; 1-11.

Avissar N, Finkelstein JN, Horowitz S. Extracellular glutathione peroxidase in human lung epithelial lining fluid and in lung cells. *Am J Physiol* 1996;270(2 Pt 1): L173-82.

Baskaran A, Chua KH, Sabaratnam V, Ravishankar Ram M, Kuppusamy UR. *Pleurotus giganteus* (Berk. Karun & Hyde), the giant oyster mushroom inhibits NO production in LPS/H2O2 stimulated RAW 264.7 cells via STAT 3 and COX-2 pathways. *BMC Complement Altern Med.* 2017;17(1):40.

Benedetti S, Benvenuti F, Pagliarani S, Francogli S, Scoglio S, Canestrari F. Antioxidant property of a novel phycocyanin extract from the blue green alga *Aphanizomenonflos-aquae. Life Sci.*2004; 24:75(19):2353-62.

Bhanja T, Kumari A, Banerjee R. Enrichment of phenolics and free radical scavenging property of wheat koji prepared with two filamentous fungi. *Bioresource Technol* 2009; 100(11):2861-6.

Borderes J, Costa A, Guedes A, Ballod Tavares LB. Antioxidant activity of the extracts from *Pycnoporussanguineus* mycelium. *Brazilian Arch Biol Technol* 2011;54(6): 1167-1174.

Borowitzka, MA. Pharmaceuticals and agrochemicals from microalgae. In: Z. Cohen, eds. *Chemicals from microalgae*, Taylor & Francis, London, UK, 1999; 313-352.

Chang HC, Yang HL, Pan JH, Korivi M, Pan JY, Hsieh MC, Chao PM, Huang PJ, Tsai CT, Hseu YC. *Hericium erinaceus* Inhibits TNF-α-Induced Angiogenesis and ROS Generation through Suppression of MMP-9/NF-κB Signaling and Activation of Nrf2-Mediated Antioxidant Genes in Human EA.hy926 Endothelial Cells. *Oxid Med Cell Longev.* 2016;8257238.

Chaurasia N, Mishra Y, Rai LC. Cloning expression and analysis of phytochelatin synthase (pcs) gene from *Anabaena* sp. PCC 7120 offering multiple stress tolerance in *Escherichia coli*. *BiochemBiophys Res Commun* 2008; 376: 225–230.

Chen Y, Zhang H, Wang Y, Nie S, Li C, Xie M· Sulfated modification of the polysaccharides from *Ganoderma atrum* and their antioxidant and immunomodulating activities. *Food Chem.* 2015;186:231-8.

Chkhikvishvili ID, Ramazanov ZM. Phenolic substances of brown algae and their antioxidant activity. *App Biochem Microbiol.* 2000; 36: 289–291.

Cilerdzic J, Stajic M, Zivkovic L, Vukojevic J, Bajic V, Spremo-Potparevic B.Genoprotective Capacity of Alternatively Cultivated Lingzhi or Reishi Medicinal Mushroom, *Ganoderma lucidum* (Agaricomycetes), Basidiocarps. *Int J Med Mushrooms.* 2016;18(12):1061-1069.

Cohen, Z., (1999). *Chemicals from microalgae*, Taylor & Francis, London, 1999. 387-489.

Comhair, SA, Erzurum SC. Antioxidant responses to oxidant-mediated lung diseases.*Am J Physiol Lung Cell Mol Physiol* 2002; 283(2):L246-55.

Cox S, Abu-Ghannam N, Gupta S. An assessment of the antioxidant and antimicrobial activity of sixspecies of edible Irish seaweeds. *Int Food Res Journal* 2010; 17, 205-220.

Cretu R, Dima S, Bahrim G. Antioxidant activity evaluation of the fungal dye oil-in-water emulsions obtained by cyclic voltammetric technique. *Romanian Biotechnol Letters* 2011;16 (6): 19-33.

Das AR, Borthakur M, Saha AK, Joshi SR, Das P. Molecular Characterization and Antioxidant Potential of Three Wild Culinary-Medicinal Mushrooms from Tripura, Northeast India. *Int J Med Mushrooms.* 2017, 19(1), 55-63.

Demirel Z, Yilmaz-koz FF, Karabay-Yavasoglu U, Ozdermir G, Sukatar A. Antimicrobial and antioxidant activity of brown algae from the Aegean Sea. *J Serb Chem Soc*. 2010; 74 (6): 619–628.

Ding Q, Yang D, Zhang W, Lu Y, Zhang M, Wang L, Li X, Zhou L, Wu Q, Pan W, Chen Y. Antioxidant and anti-aging activities of the polysaccharide TLH-3 from *Tricholoma lobayense*. *Int J Biol Macromol.* 2016;85:133-40.

Du XN, Dai JX. Screening and identification of antioxidant endophytes from *Lycium barbarum* of Ningxia. *Zhongguo Zhong Yao Za Zhi.* 2015;40(20):3941-4.

Dua A, Chauhan K, Pathak H. Biotransformation of Indigo Pigment by Indigenously Isolated *Pseudomonas* sp. HAV-1 and Assessment of Its Antioxidant Property. *Biotechnol Res Int*. 2014;109249.

Ekin S, Uzun Y, Demirel K, Bayramoglu M, Kiziltas H. Chemical Composition and Antioxidant Activity of Two Wild Edible Mushrooms from Turkey. *Int J Med Mushrooms*. 2015;17(12):1179-88.

Fahmy SE, Ali GM, Ragab MH, Ali EE, Ismaiel, MS. Biochemical composition and antioxidant activities of *Arthrospira* (*Spirulina*) *platensis* in response to gamma irradiation *Food chemistry* 2017; 214:550-555.

Fernandes MRV, Silva TAC, Pfenning LHP, etal. Biological activities of the fermentation extract of the endophytic fungus *Alternaria alternate* isolated from *Coffea Arabica* L. *Brazilian J Pharmaceutical Sci*. 2009; 45 (4): 677-685.

Franceschini S, Ceci P, Alaleona F, Chiancone E, Ilari A. Antioxidant Dps protein from the thermophilic cyanobacterium *Thermosynechococcus elongates*. *FEBS J* 2006; 273: 4913–4928.

Giridharan VV, Thandavarayan RA, Konishi T. Amelioration of scopolamine induced cognitive dysfunction and oxidative stress by *Inonotus obliquus* - a medicinal mushroom. *Food Funct*. 2011; 2(6): 320-7.

Gogavekar SS, Rokade SA, Ranveer RC, Ghosh JS, Kalyani DC, Sahoo AK. Important nutritional constituents, flavour components, antioxidant and antibacterial properties of Pleurotus sajor-caju. *J Food Sci Technol*. 2014 Aug;51(8):1483-91.

Grima EM, Belarbi EH, Fernández FG, Medina AR, Chisti YY. Recovery of microalgal biomass and metabolites: Process options and economics. *Biotechnol Adv* 2003; 20: 491-515.

Guo B. Cytonic acids A & B: novel tridepside inhibitors of hCMV protease from the endophytic fungus *Cytonaema*species. *J Nat Prod* 2000; 63: 602–604.

Gutteridge JM. Iron and oxygen a biologically damaging mixture. *Acta Paediatr Scand Suppl* 1989; 36, 78-85.

Gutteridge JMC, Halliwell B. Antioxidants: elixirs of life or media hype? *Antioxidants in nutrition, health, and disease*. Oxford University Press, Oxford 1994; 40-62.

Haana H, El-Baky A, Hussein MM, El-Baroty GS. Algal extracts improve antioxidant defense abilities and salt tolerance of wheat plant irrigated with sea water. *African J Biochem Res*2008; 2 (7): 151 –164.

Halliwell B, Murcia MA, Chirico S, Aruoma OI. Free radicals and antioxidants in food and in vivo: what they do and how they work. *Crit Rev Food Sci Nutr* 1995; 35: (1-2), 7-20.

Hao L, Sheng Z, Lu J, Tao R, Jia S.Characterization and antioxidant activities of extracellular and intracellular polysaccharides from *Fomitopsis pinicola*. *Carbohydr Polym.* 2016;141:54-9.

Henricks PA, Nijkamp FP. Reactive oxygen species as mediators in asthma. *Pulmonary Pharmacol Ther* 2001;14: 6, 409-20.

Hou Y, Ding X, Hou W. Composition and antioxidantactivity of water-soluble oligosaccharides from Hericium erinaceus. *Mol Med Rep*. 2015 May;11(5):3794-9.

Huang D, Ou B, Prior RL. The chemistry behind antioxidant capacity assays. *J Agric Food Chem* 2005; 53: 1841-56.

Huang H, Cai YZ, Xing J, Corke H, Sun M. A potential antioxidant resource: Endophytic fungi from medicinal plants.*Biomed Life Sci.*2007; 61(1): 14-30.

Huang HL, Wang BG. Antioxidant capacity and lipophilic content of seaweeds collected from the Qingdao coastline. *J Agri Food Chem* 2004; 52:4993–4997.

Ito N, Fukushima S, Tsuda H. Carcinogenicity and modification of the carcinogenic response by BHA, BHT, and other antioxidants. *CRC Crit. Rev. Toxicol* 1985; 15, 109-150.

Jan NU, Ahmad B, Ali S, Adhikari A, Ali A, Jahan A, Ali A, Ali H. Steroidal Alkaloids as an Emerging Therapeutic Alternative for Investigation of Their Immunosuppressive and Hepatoprotective Potential. *Front Pharmacol*. 2017 Mar 21;8:114.

Janjušević L, Karaman M, Šibul F, Tommonaro G, Iodice C, Jakovljević D, Pejin B. The lignicolous fungus Trametes versicolor (L.) Lloyd (1920): a promising natural source of antiradical and AChE inhibitory agents. *J Enzyme Inhib Med Chem*. 2017 Dec;32(1):355-362.

Jayanthi G, Kamalraj S, Karthikeyan K, Muthumary J. Antimicrobial and antioxidant activity of the endophytic fungus *Phomopsis* sp. GJJM07 isolated from *Mesua ferrea*. *Int J Curr Sci*. 2011; 1, 85-90.

Jia X, Zheng K, Liu S, Xu C. Optimization, Purification, Characterization, and Antioxidant Activity of Exopolysaccharide Produced by the Northern Tooth Mushroom, *Climacodon septentrionalis* (Basidiomycota). *Int J Med Mushrooms.* 2015;17(9):857-66.

Jiang L, Li Y, Wang WJ, Imoulan A, Yao YJ. Total Phenolic Content and AntioxidantActivity of Mycelial Extracts from the Medicinal Fungus *Paecilomyces hepiali* (Ascomycetes). *Int J Med Mushrooms*. 2017;19(1):35-44.

Joseph J, Panicker, SN, Janardhanan KK. Protective effect of polysaccharide-protein complex from a polypore mushroom, *Phellinus rimosus* against radiation-induced oxidative stress. *Redo Rep.* 2012; 17(1): 22-7.

Kapila S, Vibha, Sinha PR. Antioxidative and hypocholestrolemic effects of *Lactobacillus casei* ssp casei (biodefensive properties of lactobacilli). *Indian J Med Sci* 2006; 60: 361-370.

Kasuga A, Aoyagi Y, Sugaharav T. Antioxidant Activity of Fungus *Suihs bovinus* (L: Fr.) O. Kuntze. *J Food Sci*1995; 60(5), 1113-1115.

Khadhri A, Aouadhi C, Aschi-Smiti S.Screening of Bioactive Compounds of Medicinal Mushrooms Collected on Tunisian Territory. *Int J Med Mushrooms*. 2017;19(2):127-135.

Kim WS, Choi WJ, Lee S, Kim WJ, Lee DC, Sohn UD, Shin HS, Kim W· Anti-inflammatory, Antioxidant and Antimicrobial Effects of Artemisinin Extracts from *Artemisia annua* L. *Korean J Physiol Pharmacol.* 2015;19(1):21-7.

Kirkham P, Rahman I. Oxidative stress in asthma and COPD: antioxidants as a therapeutic strategy. *Pharmacol Ther* 2006; 111: 2, 476-94.

Kolesnikova LI, Kolesnikov SI, Romanova ED, Chkhenkeli VA, Darenskaya MA, Grebenkina LA, Korytov LI, Bugun OV, Koroleva NV, Gutnik IN, Antonenko FF.Effect of Preparation Based on *Trametes pubescens* Xylotroph Fungi on Lipid Peroxidation in the Blood of Experimental Animals under Conditions of Dark Stress. *Bull Exp Biol Med.* 2017;162(6):762-764.

Lea CH. Food Investigation Special Report No. 46, *Dept. Sci. Ind. Res.* His Majesty's Stationery Office, 1938.

Li P, Sun W, Luo C, Shan T, Mou Y, Lu S, Mao Z,Zhou L. *In vitro* evaluation of antioxidant activities of polysaccharides from the endophytic fungus *Berkleasmium* sp. Dzf12. *African J Microbiol Res.* 2011; 6(2): 471-477.

Lin MY, Chang FJ. Antioxidative effect of intestinal bacteria *Bifidobacteriumlongum* ATCC15708 and *Lactobacillus acidophilus* ATCC4356. *Dig Dis Sci.* 2000; 45(8):1617-22.

Liu F, Wang Y, Zhang K, Wang Y, Zhou R, Zeng Y, Han Y, Ng TB. A novel polysaccharide with antioxidant, HIV protease inhibiting and HIV integrase inhibiting activities from *Fomitiporia punctata* (P. karst.) murrill (Basidiomycota, hymenochaetales). *Int J Biol Macromol.* 2017;97:339-347.

Liu Q, Zhu M, Geng X, Wang H, Ng TB. Characterization of Polysaccharides with Antioxidant and Hepatoprotective Activities from the Edible Mushroom *Oudemansiella radicata. Molecules.* 2017: 4;22(2).

Lu XL, Yao XL, Liu Z, et al., Protective effects of xyloketal B against MPP+-induced neurotoxicity in *Caenorhabditis elegans* and PC12 cells. *Brain Res.* 2010; 1332:110-119.

Lygoe KA, Wall I, Stephens P. Role of vitronectin and fibronectin receptors in oral mucosal and dermal myofibroblast differentiation. *Biol Cell* 2007; 99: 601–614.

Maeura Y, Weisburger JH, Williams G. Dose dependent reduction of N-$_2$ fluorenylacetamide-induced liver cancer and enhancement of bladder cancer in rats by butylated hydroxytoluene. *Cancer Res* 1984; 44: 1604-08.

Maity K, Mandal E, Maity S, et al., Chemical analysis and study of immunoenhancing and antioxidant property of a glucan isolated from an alkaline extract of a somatic hybrid mushroom of *Pleurotus florida* and *Calocybe indica* variety APK2. *Int J Biol Macromol* 2012; 49(4): 555-60.

Mal'ucka LU, Harvanova J, Pavlik M, Rajtar M, Jarosciak L. Free-Radical Scavenging Activities of Cultured Mycelia of *Paecilomyces hepiali* (Ascomycetes) Extracts and Structural Characterization of Bioactive Components by Nuclear Magnetic Resonance Spectroscopy. *Int J Med Mushrooms*. 2016;18(10):895-903.

Marson Ascêncio PG, Ascêncio SD, Aguiar AA, Fiorini A, Pimenta RS. Chemical Assessment and Antimicrobial and Antioxidant Activities of Endophytic Fungi Extracts Isolated from Costus spiralis (Jacq.) Roscoe (Costaceae). *Evid Based Complement Alternat Med.*2014;2014:190543.

Masalu RJ, Hosea KM, Malendeja S. Free radical scavenging activity of some fungi indigenous to Tanzania. *Tanzania J Health Res*. 2012; 14(1); 1-8.

McVean M, Liebler DC. Prevention of DNA photodamage by Vitamin E compounds and sunscreens: roles of ultraviolet absorbance and cellular uptake. *Mol Carcinog* 1999; 24: 169-176.

Mendes RL, Nobre BP, Cardoso MT, Pereira AP, Palabra AF. Supercritical carbon dioxide extraction of compounds with pharmaceutical importance from microalgae. *Inorganica Chimica Acta* 2003; 356: 328-334.

Mendi A, Aslım B.Antioxidant lactobacilli could protect gingival fibroblasts against hydrogen peroxide: a preliminary in vitro study. *Probiotics Antimicrob Proteins*. 2014; 6(3-4):157-64.

Merghni A, Dallel I, Noumi E, Kadmi Y, Hentati H, Tobji S, Ben Amor A, Mastouri M. Antioxidant and antiproliferative potential of biosurfactants isolated from *Lactobacillus casei* and their anti-biofilm effect in oral *Staphylococcus aureus* strains.*Microb Pathog* 2017;104:84-89.

Michalak I, Chojnacka K, Saeid A, Plant Growth Biostimulants, Dietary Feed Supplements and Cosmetics Formulated with Supercritical CO_2 Algal Extracts. *Molecules* 2017; 22: 66.

Mikulikova L, Popes P. Oxidative stress, metabolism of ethanol and alcohol-related diseases. *J. Biomed. Sci.* 2001; 8: 59–70.

Mishra Y, Chaurasia N, Rai LC. AhpC (alkyl hydroperoxide reductase) from *Anabaena* sp. PCC 7120 protects *Escherichia coli* from multiple abiotic stresses. *Biochem Biophys Res Commun* 2009; 381 606–611.

Miyake Y, Ito C, Itoigawa M, Osawa T. Antioxidants produced by *Eurotium herbariorum* of filamentous fungi used for the manufacture of karebushi, dried bonito (Katsuobushi). *Biosci Biotechnol Biochem.* 2009; 73(6): 1323-1327.

Mohsin M, Negi P, Ahmed Z. Determination of the antioxidant activity and polyphenol contents of wild Lingzhi or Reishi medicinal mushroom, *Ganoderma lucidum* (W.Curt. Fr.) P. Karst. (higher Basidiomycetes) from central Himalayan hills of India. *Int J Med Mushrooms.* 2011;13(6):535-44.

Murali M, Mahendra C, Hema P, Rajashekar N, Nataraju A, Sudarshana MS, Amruthesh KN. Molecular profiling and bioactive potential of an endophytic fungus Aspergillus sulphureus isolated from Sida acuta: a medicinal plant. *Pharm Biol.* 2017 Dec;55(1):1623-1630.

Naga Satya A, Thajuddin N. antioxidant propery6 of hypersaline cyanobacteria, *Phormidium tnue* (KMD 33). *Int J Pharmacol.* 2008; 1-5.

Nakagawa H, Shiozaki T, Kobatake E, Hosoya T, Moriya T, Sakai F, Taru H, Miyazaki T. Effects and mechanisms of prolongevity induced by *Lactobacillus gasseri* SBT2055 in *Caenorhabditis elegans*. *Aging Cell.* 2016;15(2):227-36.

Narayan OP, Kumari N, Bhargava P, Rajaram H, Rai LC. A single gene all3940 (Dps) overexpression in *Anabaena* sp. PCC 7120 confers multiple abiotic stress tolerance via proteomic alterations. *Funct Integr Genomics* 2016; 16:67–78.

Narayan OP, Kumari N, Rai LC. Heterologous expression of *Anabaena* PCC 7120 all 3940 (a Dps family gene) protects *Escherichia coli* from nutrient limitation and abiotic stresses. *Biochem Biophys Res Commun.* 2010; 394(1): 163-169.

Narayan OP, Kumari N, Rai LC. Iron Starvation-Induced Proteomic Changes in *Anabaena* (Nostoc) sp. PCC 7120: Exploring Survival Strategy. *J Microbiol Bioetchnol* 2011; 21; 136-146.

Nath A, Raghunatha P, Joshi SR. Diversity and biological activities of endophytic fungi of *Emblica officinalis*, an ethnomedicinal plant of India. *Mycobiol.* 2012; 40(1): 8-13.

Novakovic A, Karaman M, Kaisarevic S, Radusin T, Llic N. Antioxidant and Antiproliferative Potential of Fruiting Bodies of the Wild-Growing King Bolete Mushroom, *Boletus edulis* (Agaricomycetes), from Western Serbia. *Int J Med Mushrooms.* 2017;19(1):27-34.

Öztürk M, Duru ME, Kivrak S, Mercan-Doğan N, Türkoglu A, Özle, MA. *In vitro* antioxidant, anticholinesterase and antimicrobial activity studies on three *Agaricus* species with fatty acid compositions and iron contents: a comparative study on the three most edible mushrooms. *Food Chem Toxicol.* 2011; 49(6): 1353-60.

Pandey S, Rai R, Rai LC. Proteomics combines morphological, physiological and biochemical attributes to unravel the survival strategy of *Anabaena* sp. PCC7120 under arsenic stress. *J Proteomics.* 2012; 75;921-937.

Peña MMO, Burkhart W, Bullerjahn GS. Purification and characterization of a *Synechococcus* sp. strain PCC 7942 polypeptide structurally similar to the stress-induced Dps/PexB protein of *Escherichia coli, Arch. Microbiol.* 1995; 163:337–344.

Qiu J, Wang X, Song C. Neuroprotective and antioxidant lanostanoid triterpenes from the fruiting bodies of *Ganoderma atrum. Fitoterapia.* 2016;109:75-9.

Raghava Rao KV, Mani P, Satyanarayana B, Raghava Rao T. Purification and structural elucidation of three bioactive compounds isolated from *Streptomyces coelicoflavus* BC 01 and their biological activity. *Biotech.* 2017;7(1):24.

Rahman I, Biswas SK, Kodc A. Oxidant and antioxidant balance in the airways and airway diseases. *Eur J Pharmacol* 2006; 533: 222–239.

Rani P, Lal MR, Maheshwari U, Krishnan S. Antioxidant Potential of Lingzhi or Reishi Medicinal Mushroom, *Ganoderma lucidum* (Higher Basidiomycetes) Cultivated on *Artocarpus heterophyllus* Sawdust Substrate in India. *Int J Med Mushrooms.* 2015;17(12):1171-7.

Rani RP, Anandharaj M, Sabhapathy P, Ravindran AD. Physiochemical and biological characterization of novel exopolysaccharide produced by *Bacillus tequilensis* FR9 isolated from chicken. *Int J Biol Macromol.* 2017; 96:1-10.

Rodrigues KF, Costa,GL, Carvalho MP, Epifanio RDA. Evaluation of extracts produced by some tropical fungi as potential cholinesterase inhibitors. *World J Microbiol Biotechnol.*2005; 21(8-9):1617–1621.

Romay C, Armesto J, Remirez D, Gonza'lez R, Ledon N, Garcia I. Antioxidant and anti-inflammatory properties of C-phycocyanin from blue-green algae. *Inflamm Res.*1998; 47: 36–41.

Sarikurkcu C, Tepe B, Kocak MS, Uren MC. Metal concentration and antioxidantactivity of edible mushrooms from Turkey. *Food Chem.* 2015;175:549-55.

Shappirio, S., *US patent no.* 2338207 (1944).

Sharma SK, Gautam N, Atri NS. Optimized extraction, composition, antioxidant and antimicrobial activities of exo and intracellular polysaccharides from submerged culture of *Cordyceps cicadae. BMC Complement Altern Med.* 2015;15:446.

Sharma SK, Gautam N.Chemical Composition and Antioxidant and Antibacterial Activities of Cultured Mycelia of Four Clavicipitaceous Mushrooms (Ascomycetes) from the Indian Himalayas. *Int J Med Mushrooms.* 2017;19(1):45-54.

Shashidhar MG, Manohar B. Acidified Hot Water Extraction of Adenosine, Cordycepin, and Polysaccharides from the Chinese Caterpillar Mushroom, *Ophiocordyceps sinensis* CS1197 (Ascomycetes): Application of an Artificial Neural Network and Evaluation of Antioxidant and Antibacterial Activities. *Int J Med Mushrooms.* 2016;18(10):915-926.

Simontacchi M, Caro A, Fraga CG, Puntarulo S. Oxidative stress affects α-tocopherol content in soyabean embryonic axes upon inhibition and following germination. *Plant Physiol* 1993;103: 949-953.

Smith JL, Alford JA. Presence of antioxidant materials in bacteria. *Lipids* 1970;5(10):795-9.

Srinivasan K, Jagadish LK, Shenbhagaraman R, Muthumary J. Antioxidant activity of endophytic fungus Phyllosticta sp. isolated from *Guazuma tomentosa*. *J Phytology Phytochem*. 2010; 2(6):37–41.

Strobel G, Yang X, Sears J, Kramer R, Sidhu RS, Hess WM. Taxol from *Pestalotiopsis microspora*, an endophytic fungus of *Taxus wallichiana*. *Microbiol.* 1996; 142: 435–440.

Sun C, Wang JW, Fang L, Gao XD, Tan RX. Free radical scavenging and antioxidant activities of EPS2, an exopolysaccharide produced by marine filamentous fungus *Keissleriella* sp. YS 4108. *Life Sci*2004; 75(9): 1063-73.

Surekha Ch, Kaladhar DSVGK, Raju SJR. Haseena. Evaluation of antioxidant and antimicrobial potentiality of some edible mushrooms. *Int J Adv Biotechnol Res* 2011; 2:130-134.

Suseem SR, Saral M. Inhibition of platelet aggregation and in vitro free radical scavenging activity of dried fruiting bodies of *Pleurotus eous*. *Chin J Integr Med.* 2015 Jul;21(7):530-6.

Takao T, Kitatani F, Watanabe N, Yagi A, Sakata K. A simple screening for antioxidants and isolation of several antioxidants produced by marine bacteria from fish and shelfish. *Biosci Biotech Biochem* 1994; 58 (10): 1780-83.

Tan RX, Zou WX. Endophytes: a rich source of functional metabolites. *Nat Prod Reports* 2001; 18:448–459.

Tang W, Xing Z, Li C, Wang J, Wang Y. Molecular mechanisms and in vitro antioxidant effects of *Lactobacillus plantarum* MA2. *Food Chem.* 2017;221:1642-1649.

Tosin AO, Ademola OO, Anthony IO. *Therapeutic Potentials of Microalgae in the Treatment of Alzheimer's Disease Molecules* 2017; 22; 480.

Vijayakumar M, Ilavenil S, Kim DH, Arasu MV, Priya K, Choi KC. In-vitro assessment of the probiotic potential of *Lactobacillus plantarum* KCC-24 isolated from Italian rye-grass (Lolium multiflorum) forage. *Anaerobe.* 2015;32:90-7.

Vunduk J, Klaus A, Kozarski M, Petrovic P, Zizak Z, Niksic M, Van Griensven LJ. Did the Iceman Know Better? Screening of the Medicinal Properties of the Birch Polypore Medicinal Mushroom, *Piptoporus betulinus* (Higher Basidiomycetes). *Int J Med Mushrooms.*2015;17(12):1113-25.

Wang J, Liu YM, Cao W, Yao KW, Liu ZQ, Guo JY. Anti-inflammation and antioxidant effect of Cordymin, a peptide purified from the medicinal mushroom *Cordyceps sinensis*, in middle cerebral artery occlusion-induced focal cerebral ischemia in rats. *Metab Brain Dis* 2012; Feb: 12.

Wang ZB, Pei JJ, Ma HL, Cai PF, Yan JK. Effect of extraction media on preliminary characterizations and antioxidant activities of *Phellinus linteus* polysaccharides. *Carbohydr Polym.* 2014;109:49-55.

Wang ZR, Sheng JP, Tian XL, Wu TT, Liu WZ, Shen L, The in vitro antioxidant properties of *Bacillus simplex* XJ25 isolated from sand biological crusts. *African J Microbiol Res*. 2011; 5 (28):4980-4986.

Wei,W, Luo X, Zheng L, Yu M, Jiang N, Xu XY, Yang ZR. Isolation of a wild *Morchella* spp. strain and the effects of its extract on ethanol-induced gastric mucosal lesions in rats. *Z Naturforsch C*. 2011; 66(1-2): 55-62.

Wu M, Luo X, Xu X, Wei W, Yu M, Jiang N, Ye L, Yang Z, Fei X. Antioxidant and immunomodulatory activities of a polysaccharide from *Flammulina velutipes*. *J Tradit Chin Med*. 2014; 34(6):733-40.

Xia Z. Preparation of the oligosaccharides derived from *Flammulina velutipes* and their antioxidant activities. *Carbohydr Polym*. 2015;118:41-3.

Xie P, Fujii I, Zhao J, Shinohara M, Matsukura M. A novel polysaccharide derived from algae extract induces apoptosis and cell cycle arrest in human gastric carcinoma MKN45 cells via ROS/JNK signaling pathway. *Int. J. Oncol*. 2016; 49: 1561–1568.

Xu N, Ren Z, Zhang J, Song X, Gao Z, Jing H, Li S, Wang S, Jia L. Antioxidant and anti-hyperlipidemic effects of mycelia zinc polysaccharides by *Pleurotus eryngii* var. tuoliensis. *Int J Biol Macromol*. 2017;95:204-214.

Yang Y, Hu J, Liu, Y, et al. Antioxidant and cytotoxic activities of ethanolic extracts and isolated fractions of species of the genus *Phellinus Quél*. (Aphyllophoromycetideae). *Int J Med Mushrooms* 2011, 13(2):145-52.

Yen GC, Chang YC, Sheu F, Chiang HC. Isolation and Characterization of Antioxidant Compounds from *Aspergillus candidus* Broth Filtrate. *J Agric Food Chem* 2001;49(3):1426–1431.

Yim HS, Chye FY, Liow ML, Ho CW. Antioxidant Potential of *Pleurotus porrigens* Extract and Application in Sunflower Oil during Accelerated Storage. *Chiang Mai J. Sci*. 2013; 40(1): 34-48.

Yin X, You Q, Zhou X. Complex Enzyme-Assisted Extraction, Purification, an AntioxidantActivity of Polysaccharides from the Button Mushroom, *Agaricus bisporus* (Higher Basidiomycetes). *Int J Med Mushrooms*. 2015;17(10):987-96.

Yoon KN, Alam N, Lee KR, Shin PG, Cheong JC, Yoo YB, Lee TS. Antioxidant and antityrosinase activities of various extracts from the fruiting bodies of *Lentinus lepideus*. *Molecules*. 2011;16 (3): 2334-47.

Yuan YV, Walsch NA. Antioxidant and antiproliferative activities of extracts from a variety of edible seaweeds. *Food Chem Toxicol*. 2006; 44(7): 1144-50.

Yurkiv B, Wasser SP, Nevo E, Sybirna NO. Antioxidant Effects of Medicinal Mushrooms *Agaricus brasiliensis* and *Ganoderma lucidum* (Higher Basidiomycetes): Evidence from Animal Studies. *Int J Med Mushrooms*. 2015;17(10):943-55.

Zahra R, Mehrnaz M, Farzaneh V, Sartavi Kohzad S. Antioxidant activity of extract from a brown alga, *Sargassum boveanum*. *African J Biotechnol* 2007; 6 (24), 2740-2745.

Zeng PY, Wu JG, Liao LM, Chen TQ, Wu JZ, Wong KH. In vitro antioxidant activities of endophytic fungiisolatedfrom the liverwort *Scapania verrucosa*. *Genet Mol Res* 2011;10(4): 3169-79.

Zhang J, Gao Z, Li S, Gao S, Ren Z, Jing H, Jia L. Purification, Characterization, Antioxidation, and Antiaging Properties of Exopolysaccharides and Endopolysaccharides of the Royal Sun Medicinal Mushroom, *Agaricus brasiliensis* (Agaricomycetes). *Int J Med Mushrooms*. 2016;18(12):1071-1081.

Zhang JC, Chen GY, Li XZ, Hu M, Wang BY, Ruan BH, Zhou H, Zhao LX, Zhou J, Ding ZT, Yang YB. Phytotoxic, antibacterial, and antioxidant activities of mycotoxins and other metabolites from *Trichoderma* sp. *Nat Prod Res*. 2017 Mar 2:1-8.

Zheng LP, Gao LW, Zhou JQ, Sima YH, Wang JW. Antioxidant activity of aqueous extract of a *Tolypocladium* sp. fungus isolated from wild *Cordyceps sinensis*. *African J Biotechnol*2008; 7 (17): 3004-3010.

In: Recent Advances in Microbiology. Volume 3 ISBN: 978-1-53614-057-6
Editors: N.C. Gautam, S.P. Tiwari and R. Sharma

Chapter 3

PRODUCTION OF NATURAL VANILLA FROM FERULIC ACID USING BIOTRANSFORMATION TOOLS

Shashwati Ghosh Sachan*[1,*]*, Ashish Sachan*[1]*, Adinpunya Mitra*[2]*,
***Shashank Mishra*[1]*, Meenakshi Kullu*[1]* and A. S. Vidyarthi*[1]**
[1]Department of Biotechnology, Birla Institute of Technology, Mesra, Ranchi, India
[2]Natural Product Biotechnology Group, Agricultural and Food Engineering Department, Indian Institute of Technology Kharagpur, Kharagpur, India

NATURAL VANILLA

Vanilla is the world's most popular flavoring agent for sweetened foods. Pure vanilla extracts are widely used for flavoring in ice creams, chocolates, tobacco, baked foods, puddings, cakes, liquors and as a fragrance ingredient in perfumery. It improves the flavor of caramel, coffee and some dairy products. It is also used in spice oleoresin formulations, sausages, seasonings etc. Natural vanilla is a complex mixture of flavor components extracted from the cured beans of the *Vanilla* plant (Ranadive, 1994). More than 170 volatile aromatic components have been identified, apart from the principal components vanillin, vanillic acid, *p*-hydroxybenzaldehyde, vanillyl alcohol, *p*-hydroxybenzoic acid and *p*-hydroxybenzylalcohol that are present in detectable amounts in the *Vanilla* pod (Walton et al., 2003). Natural vanillin also acts as an antimicrobial agent by inhibiting molds and yeast in fruit purees and fruit-based agar systems (Cerrutti et al., 1997).

[*] Corresponding Author E-mail:ssachan@bitmesra.ac.in.

Keywords: vanilla, ferulic acid, biotransformation

NATURAL VANILLA FLAVOR BIOSYNTHESIS IN *VANILLA* PODS

The source of natural vanilla flavor is the bean or pod of the tropical climbing orchid *Vanilla planifolia* (family Orchidaceae), cultivated for its pleasant flavor. Out of the 50 species known, only three are found to be important as sources of the natural flavor. These species are *Vanilla planifollia*, *Vanilla pompona* and *Vanilla tahitensis*. Among these, *Vanilla planifolia* is the most preferred species and is being cultivated commercially. *V. planifolia* traces its origin to Mexico, from where the Spainards brought it to Europe in 1520 where it gained popularity The major *Vanilla* producing countries are Madagascar, Indonesia, Mexico and Comoros. In India, *Vanilla* is cultivated in Kerala, Karnataka and Tamilnadu. Trace amounts of vanillin are also found in other plants like daffodil, hyacinth, potato and tobacco (Havkin-Frenkel et al., 1999). However, the pods of the *Vanilla* orchid still remain the only commercial source of natural vanilla flavor. Various factors such as plant species, growing conditions, soil nutrition, harvesting maturity and curing method are taken into consideration when determining the relative concentrations of the constituents of vanilla flavor. The cured beans responsible for flavor generation contain volatile aromatics such as carbonyls, aliphatic alcohols, lactones, aromatic and aliphatic hydrocarbons, terpenoids, heterocyclics and aromatic alcohols, acids, and esters. The non-volatile constituents that add to the characteristic vanilla flavor are tannins, polyphenols, resins and free amino acids. All these constituents together impart a delicate and rich aroma.

The vanilla aroma develops in the pods or beans through a rather labor-intensive process called curing, which is performed to dry the *Vanilla* beans and to allow chemical and enzymatic reactions to occur for aroma development. Every *Vanilla*-cultivating country has developed its own curing process, but all processes generally consist of four basic steps: scalding, sunning, drying and conditioning. In the scalding procedure, the vegetative phase is stopped by soaking the beans in hot water for a few minutes. This is followed by the sunning procedure, where the beans are spread out in the sun until they are hot, and enzymes like glucosidase and oxidases are very active. This gives the beans their characteristic brown color. Subsequently, these beans are wrapped up in sheets (sweating), kept overnight into airtight containers and the process is repeated for 1 to 2 weeks. The drying of the beans is performed indoors for a period of 2-4 weeks, which prevents them from being spoiled by microorganisms. The most time-consuming step is the conditioning process (several months), in which several biochemical reactions take place (Ranadive, 1994). Significant research has been conducted in order to speed up the curing process, and the updated literature on this has been covered in a detailed review by Dignum et al., (2001).

Demand and Commercial Importance of Natural Vanillin

Natural vanillin is the world's most highly prized natural flavor used in foods, beverages, perfumes and pharmaceuticals (Clark, 1990). The price of chemically synthesized "nature-identical" vanilla flavor is very low, compared to the price of cured vanilla pods. The price of natural vanillin extracted from vanilla pods usually varies between $1200 per kg and $4000 per kg, in contrast to the price of synthetic vanilla, which is less than $15 per kg (Lomascolo et al., 1999). The high price of natural vanilla flavor is mainly due to the limited availability of *Vanilla* pods, depending on climate-associated fluctuations of harvest yields, economic and political decisions, and last but not the least, the labor-intensive cultivation, pollination, harvesting and curing of *Vanilla* pods as described above. However, the increasing trend towards obtaining natural food ingredients has helped maintain buoyant demand for natural vanillin. Unfortunately, the worldwide demand for natural *Vanilla* beans cannot be met by the available supply from limited growing areas and the above-mentioned challenges. On account of the limited supply and high price of natural vanillin and the predominance of chemically synthesized vanillin, there arose an incentive to explore and develop biological sources of 'natural vanillin' and vanilla-type flavoring that could be marketed as a realistic alternative to the chemically synthesized substances. Biotransformation technology, utilizing either plant cells or microbes, has emerged as a viable option to ensure stable supply, quality and price (Funk and Brodelius, 1994; Westcott et al., 1994; Labuda et al., 1993; Priefert et al., 2001)

Natural vanilla is of interest to scientists for two main reasons. One concerns the relationship of vanillate derivatives to be explored, using the phenylpropanoid pathway and the mechanisms of formation of benzoic acids, including 4-hydroxybenzoic acid and salicylic acid, the signaling compound. The other relates to the commercial importance of these compounds and hence, the possibilities of producing them by biotechnology (Walton et al., 2003).

Biotechnological Methods for the Production of Natural Vanilla Flavor

Biotransformation is a broad and growing field of biotechnology, which involves the use of enzymes or living cells to perform chemical reactions, in which the starting substances and products are of comparable chemical complexity. Basically, biotransformation is a chemical reaction carried out on natural or synthetic substrates using biocatalysts. The most important features of biocatalysts are 1) they are efficient, 2) they are environmentally acceptable, 3) they operate under mild conditions, 4) they are selective and 5) they are cost-effective. A large number of natural products are generated via microbial or plant cell biotransformation where a powerful arsenal of enzymes and metabolic pathways are involved in their synthesis. Biotransformation is now used for the preparation of valuable chemicals for the pharmaceutical, agricultural, flavor and fragrance, nutritional and chemical industries (Beilen and Li, 2002; Cheetham, 1993).

PLANT TISSUE/CELL CULTURE-MEDIATED PRODUCTION OF NATURAL VANILLA FROM FERULIC ACID

Plant tissues have been shown to produce a wide range of flavors and aroma compounds, characteristic of their plant origin. The rationale for the production of flavor compounds with cultured plant cells is based upon their unique biochemical and genetic capacity and totipotency of plant cells. Thus, in principle, every cell of a *Vanilla* plant contains the genetic information necessary to produce the numerous components that constitute the flavor of natural vanillin. Feeding intermediates of the biosynthetic pathway can enhance the production of vanilla flavor metabolites through precursor biotransformation. Feeding of ferulic acid to *V. planifolia* cells resulted in an enhanced accumulation of vanillin, as compared to the untreated callus (Romagnoli and Knorr, 1988). Cinnamic acid and ferulic acid, when fed to cell suspension cultures of *Vanilla planifolia*, resulted in the formation of traces of *p*-hydroxybenzoic acid and vanillic acid (Funk and Brodelius, 1990). The influence of light on the growth of *V. planifolia* callus cultures and on the accumulation of vanillin and relative vanillin precursors was studied by Havkin-Frenkel et al., (1996). The nature of vanilla component precursors added to the medium was among other factors influencing flavor production in *V. fragrans* cultures. Westcott et al., (1994) developed a process for producing natural vanillin from ferulic acid as the precursor, with aerial roots and charcoal as a product reservoir. Biotransformation to vanillin has been reported in both freely suspended and immobilized cell cultures of *Capsicum frutescens* treated with the phenylpropanoid pathway intermediates (Johnson et al., 1996; Ramachandra and Ravishankar, 1999; Suresh et al., 2003).

PRODUCTION OF NATURAL VANILLA FROM FERULIC ACID USING MICROBES

A promising alternative to obtain natural products is the application of microbial biotransformation tools. Till date, the US and European legislatures have both defined the term 'natural flavor' as compounds obtained from living cells, or their components, including enzymes (Lomascolo et al., 1999). Since vanillin can be produced by the chemical breakdown of lignin, it is not unreasonable to believe that microbes involved in lignin degradation might possess enzymes for natural vanillin formation. Organisms capable of degrading ferulic acid have been of particular interest, since the phenolic substituents in this compound are also encountered in vanillin and vanillic acid (Figure 1).

Among various phenylpropanoid compounds, the hydroxycinnamic acids are quite amenable for microbial degradation. Microbes transform hydroxycinnamic acids (e.g., ferulic acid, *p*-coumaric acid, sinapic acid) to their corresponding hydroxybenzoates, which are important components of natural flavors and fragrances. A large number of microbes such as bacteria, fungi and yeast were explored for the laboratory-scale production of vanilla flavor metabolites from various phenylpropanoid precursors such as ferulic acid, eugenol, isoeugenol, vanillyl alcohol, and vanillylamine. (Ramachandra and Ravishankar, 2000). Of these, ferulic acid is the most promising substrate, a very common constituent of plant cell wall wastes such as wheat and oat bran, sugarbeet pulp and sugarcane bagasse. Ferulic acid comprises 0.5% (w/w) of wheat bran, 0.8% of sugar beet pulp and 3.1% of maize bran (Ou and Kwok, 2004). It is the most abundant hydroxycinnamic acid in the plant kingdom, found either esterified or etherified with the polysaccharides and lignin in the cell wall (Figure 2) (Sun et al., 2002).

The biodegradation of hydroxycinnamic acids is also important from the perspective of the global carbon cycle and from an environmental point of view, since these compounds are abundant in nature and are released from plant wastes as breakdown products from lignin (Peng et al., 2003).

There has been rapid progress in the biochemistry and molecular biology of natural vanillin formation in microorganisms. This has opened up possibilities for generating natural vanillin using different microorganisms including food-fermentative organisms. The text that follows presents an updated literature review (including patents) on microbial biotransformation of ferulic acid to natural vanilla flavor compounds using fungi, bacteria and actinomycetes.

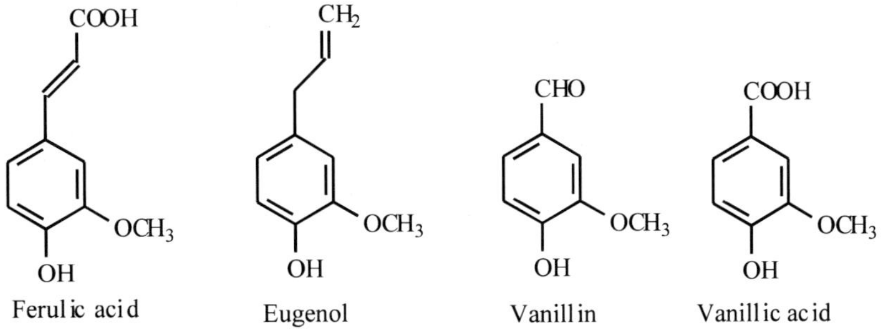

Figure 1. Identical phenolic substituents of potential substrates and anticipated products of natural vanillin.

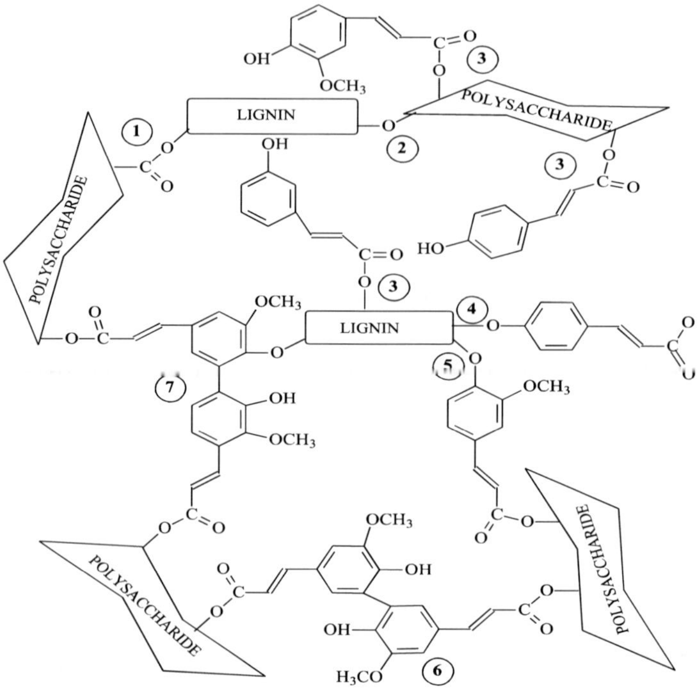

Figure 2. Position of ferulic acid in the cell wall of plants

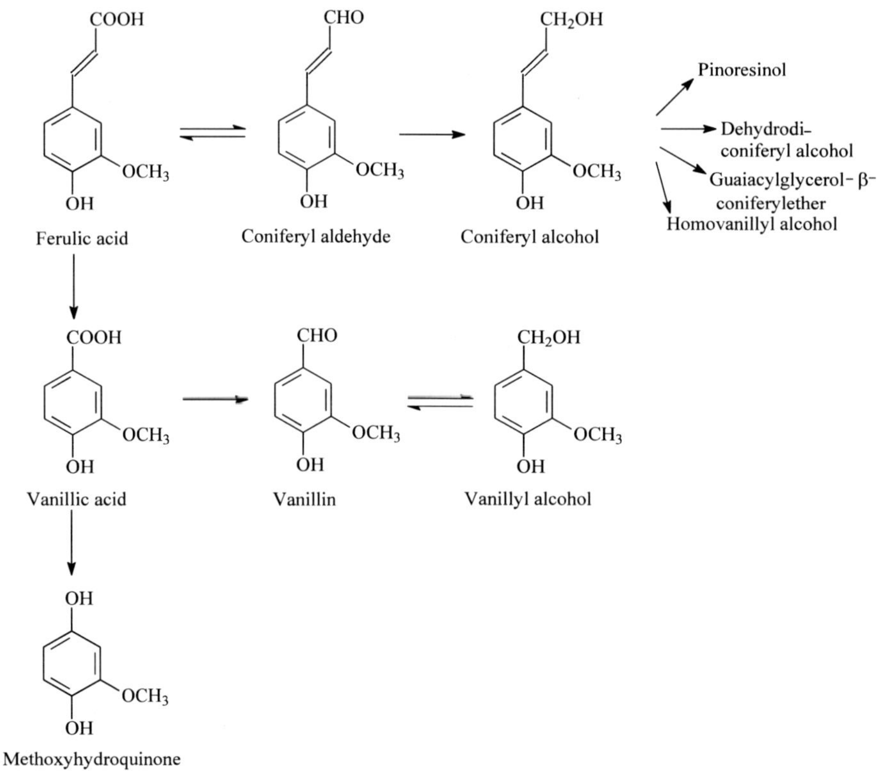

Figure 3. Metabolism of ferulic acid to vanillin by *Pycnoporus cinnabarinus*

By Fungi

Ferulic acid is a well-known product of lignin degradation by white-rot fungi. Among fungi, *Trametes* sp. reduced ferulic acid into coniferyl alcohol, which was further degraded to vanillic acid, vanillyl alcohol and methoxyhydroquinone (Nishida and Fukuzumi, 1978). The degradation of ferulic acid by *Pycnoporus cinnabarinus* has been shown to proceed via vanillic acid formation, which was subsequently reduced to vanillin, vanillyl alcohol or decarboxylated to methoxyhydroquinone (Falconnier et al., 1994) (Figure 3).

Ferulic acid metabolism has also been thoroughly studied in *Sporotrichum pulverulentum* (Ander et al., 1980; Gupta et al., 1981), which produced small amounts of vanillin, vanillic acid and methoxyhydroquinone. Another strategy involved the complementary bioconversion abilities of two filamentous fungi. In the first step, the micromycete *Aspergillus niger* transformed ferulic acid into vanillic acid, while in the second step, vanillic acid was metabolized into vanillin by the white-rot fungus, *Pycnoporus cinnabarinus* (Lesage-Meessen et al., 1996; Lesage-Meessen et al., 1997; Bonnin et al., 1999). Further, addition of cellobiose in the medium showed inhibition of the vanillic acid decarboxylation catalyzed by laccase or peroxidase (Ander et al., 1990) and thus, limited the formation of methoxyhydroquinone and favored vanillin production (Lesage-Meessen et al., 1997; Bonnin et al., 1999). The vanillin yield was further improved by the use of high-density cultures (Oddou et al., 1999). In another study, the biotechnological process of vanillin production was scaled-up at the laboratory level using two types of bioreactors (Stentelaire et al., 2000). It was found that mechanically aerated bioreactors were more efficient as compared to air-lift bioreactors. These investigations led to a two-step process combining the biotransformation of sugar-beet pulp ferulic acid to vanillic acid by *Aspergillus niger* and the biotransformation of recovered vanillic acid to vanillin by *P. cinnabarinus* (Lesage-Meessen et al., 1996).

Fusarium solani was shown to degrade ferulic acid via the formation of vinyl-guaiacol. Ferulic acid was metabolized by a *Penicillium* sp. in a novel route, involving a preliminary demethylation to caffeic acid, followed by side-chain shortening to yield protocatechuic acid that was subsequently broken down *via* the ortho-fission pathway (Tillett and Walker, 1990). Bioconversion of ferulic acid by *Sporotrichum thermophile* was studied by Topakas et al., (2003), where vanillic acid was obtained as the major degradation product. Ferulic acid metabolism by *S. thermophile* apparently occurred *via* the propenoic chain degradation, where the formation of 4-hydroxy-3-methoxystyrene (4-vinylguaiacol) was detected. This 4-vinylguaiacol was presumably metabolized to vanillic acid. Vanillin, vanillyl alcohol, vanillic acid, methoxyhydroquinone, guaiacol and catechol were detected in addition to 4-vinylguaiacol as the degradation products upon ferulic acid metabolism by *Paecilomyces variotii* and *Pestalotia palmarum* (Rahouti et al., 1989). Krings et al. (2001) studied the pathway of ferulic acid degradation by *P.*

cinnabarinus using [5-^{2}H]-labeled ferulic acid. Detection of considerable amounts of labeled 4-hydroxy-3-methoxyacetophenone in the late growth phase suggested an increase in the formation of 4-hydroxy-3-methoxybenzoylacetic acid and its subsequent decarboxylation. Thus, a β-oxidation-like degradation of ferulic acid or its methyl ester to vanillic acid was established.

In another study, *Schizophyllum commune* showed the capability of converting ferulic acid into vanillic acid with transient formation of low amounts of vanillin (Ghosh et al., 2005). Similar results were reported in the case of *Paecilomyces variotii* MTCC 6581 (Ghosh et al., 2006). Ferulic acid biotransformation by the wild strain *Aspergillus niger* C28B25 and a diploid strain DAR2, obtained by parasexual recombination, was studied by Baqueiro-Peña et al. (2010). The wild strain of *A. niger* C28B25 biotransforms ferulic acid to vanillic acid, while the diploid strain DAR2 preferentially decarboxylates ferulic acid to 4-vinylguaiacol.

By Bacteria

The biodegradation of ferulic acid has also been investigated in many bacteria. Bacterial species that have been shown to form vanillin as an intermediate in ferulic acid metabolism include *Pseudomonas acidovorans* (Toms and Wood, 1970), *Escherichia coli* (Ötuk, 1985), and *Bacillus subtilis* (Gurujeyalakshmi and Mahadevan, 1987). It was reported that *Pseudomonas cepacia* has the capacity to degrade ferulic acid via 4-vinylguaiacol to vanillin and vanillic acid (Andreoni et al., 1984). The degradation of ferulic acid was proposed to proceed via 4-hydroxy-3-methoxy phenylpropionic acid, vanillic acid and protocatechuic acid in *Pseudomonas fluorescens* FE2 (Andreoni and Bestetti, 1988). Under certain optimal conditions, *Pseudomonas fluorescens* BF13 was reported to transiently accumulate vanillic acid upon the degradation of ferulic acid (Barghini et al., 1998). A novel CoA-dependent yet non-β-oxidative pathway was demonstrated in *Pseudomonas fluorescens* AN103 (Gasson et al., 1998; Narbad and Gasson, 1998; Mitra et al., 1999) for the first time, where the microbe transformed ferulic acid to feruloyl-CoA, which subsequently underwent side chain cleavage to form vanillin and acetyl-CoA. This vanillin was subsequently converted into vanillic acid, which later underwent ring opening via the formation of protocatechuic acid (Figure 4). The same pathway was found to be operating in *Pseudomonas* sp. strain HR 199 (Overhage et al., 1999). A mutant of *P. fluorescens* BF13 was later isolated, which retained the ability to bioconvert ferulic acid into vanillic acid but was unable to degrade vanillic acid further (Civolani et al., 2000). Biodegradation of ferulic acid by *Sphingomonas paucimobilis* SYK-6 (Masai et al., 2002), *Delftia acidovorans* (Plaggenborg et al., 2001) and *Amycolatopsis* sp. strain HR167 (Achterholt et al., 2000) were also shown to follow the CoA-dependent, non- β-oxidative pathway as described above.

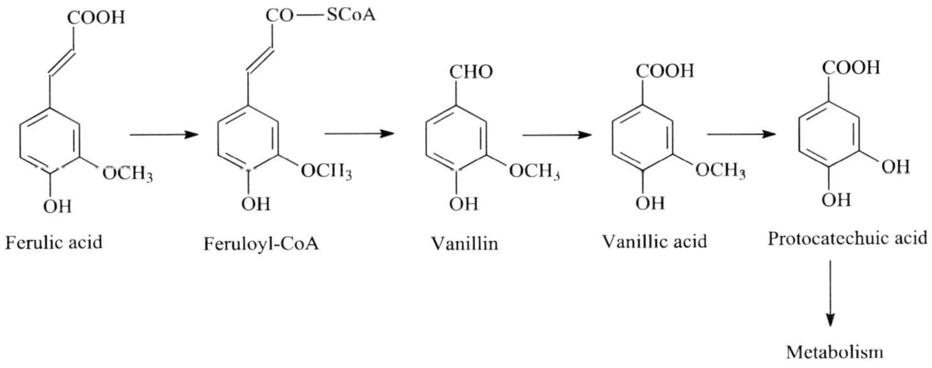

Figure 4. Ferulic acid metabolism in *Pseudomonas fluorescens* AN103

Several microorganisms such as *Rhodotorula rubra* (Huang et al., 1994), *B. coagulans* (Karmakar et al., 2000), *Pestalotia palmarum* (Rahouti et al., 1989), *B. pumilus* PS213 (Degrassi et al., 1995) and *B. subtilis* (Cavin et al., 1998) were shown to degrade ferulic acid via 4-vinylguaiacol to vanillin and then to vanillic acid. In *Corynebacterium glutamicum*, dihydroferulic acid, vanillin and vanillic acid were identified as metabolites (Labuda et al., 1993). Apart from 4-vinylguaiacol, the occurrence of dihydroferulic acid was observed in *S. cerevisiae* (Huang et al., 1993). A biotechnological production process was developed using *Bacillus pumilus*, which yielded 4-vinylguaiacol from ferulic acid (Lee et al., 1998). In the recent past, knockout experiments have revealed the existence of a second phenolic acid decarboxylase in *Lactobacillus plantarum*, which is induced better with ferulic acid than with *p*-coumaric acid (Barthelmebs et al., 2000). The degradation of ferulic acid by a thermophilic *Bacillus* (strain B1) also resulted in the formation of vanillin, vanillic acid and 4-hydroxy-3-methoxyphenyl-*β*-hydroxypropionic acid, as demonstrated by Peng et al., (2003). In *Wolinella succinogenes*, ferulic acid was reduced to 4-hydroxy-3-methoxy phenylpropionic acid, which has been further converted into vanillic acid through homovanillic acid (Ohmiya et al., 1986). *Acinetobacter calcoaceticus* DSM 586 was found to degrade ferulic acid to vanillic acid (Delneri et al., 1995). In the case of *Enterobacter* sp. Px6-4 (Li et al., 2008), the primary metabolites using ferulic acid as the sole carbon source included 4-vinylguaiacol, vanillin as well as other related chemicals, all of which shared the same guaiacol-carbon-skeleton. *Lactobacillus farciminis* ATCC 29644 also showed the potential to convert ferulic acid to vinyl guaiacol (Adamu et al., 2012).

By Actinomycetes

Among the microorganisms, actinomycetes were also shown to be efficient in biotransformation of different phenylpropenoic acids. During the metabolism of ferulic

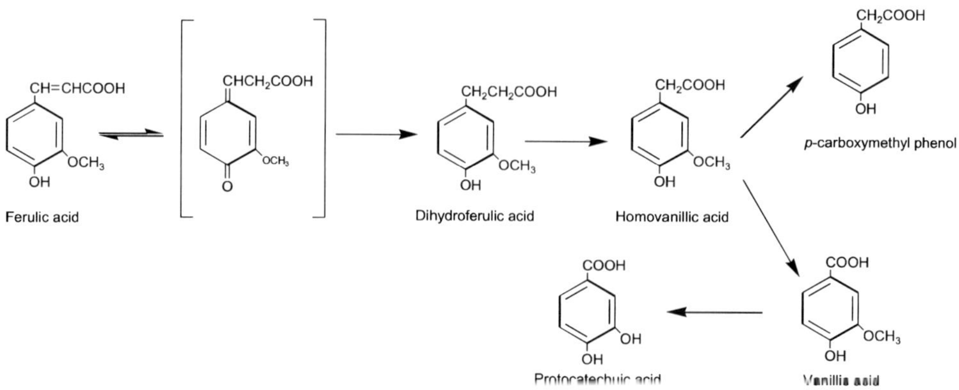

Figure 6B.

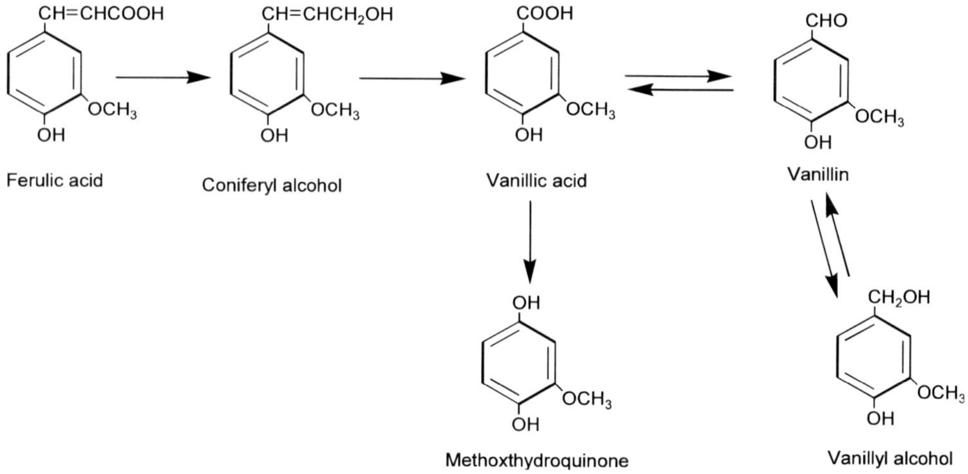

Figure 6C.

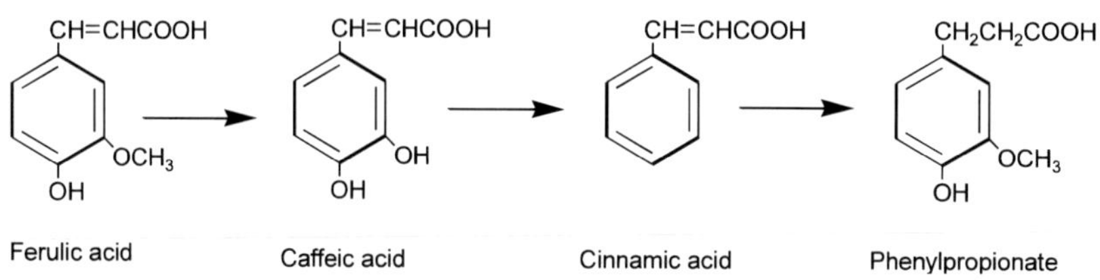

Figure 6D.

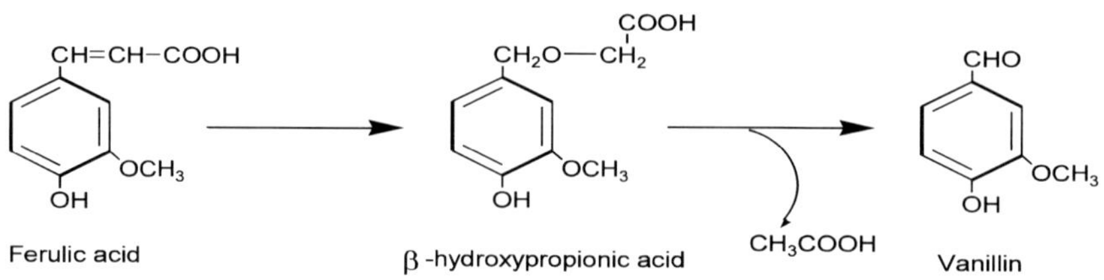

Figure 6E.

Figure 6. (Continued).

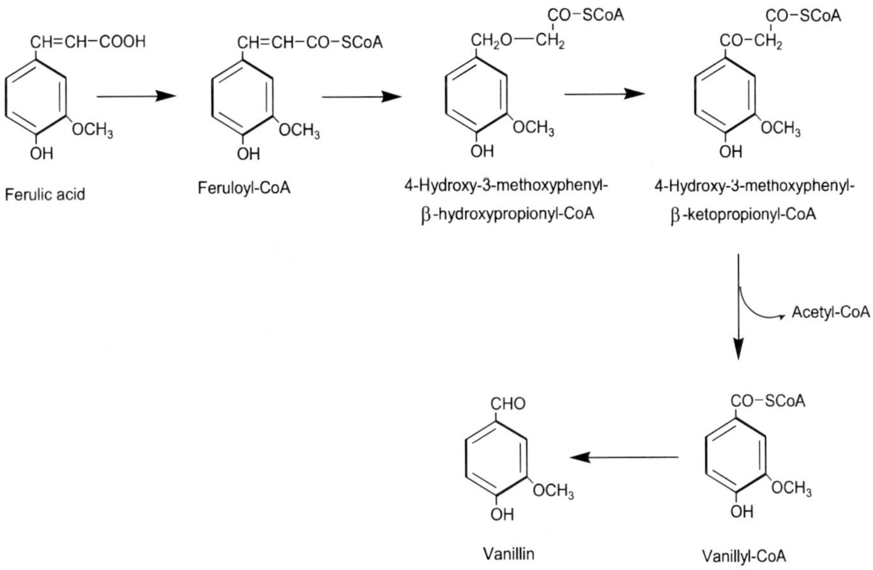

Figure 6F.

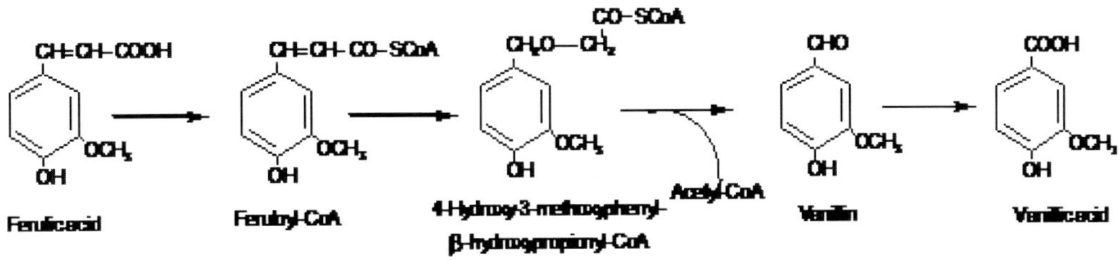

Figure 6G.

Figure 6. A-G: Major pathways of ferulic acid degradation. Decarboxylation (A), Reductive conversions (B-D), Coenzyme A- independent deacetylation (E), coenzyme A- dependent deacetylation (β- oxidation) (F) and coenzyme A- dependent deacetylation (non-β- oxidation) (G) of ferulic acid.

In the *fourth mechanism* (coenzyme-A-dependent deacetylation or *β*-oxidation), a coenzyme-A-dependent pathway analogous to the *β*-oxidation pathway of fatty acid catabolism has been proposed for the degradation of ferulic acid (Huang et al., 1994). This pathway includes the thioclastic cleavage of 4-hydroxy-3-methoxyphenyl- *β*-ketopropionyl-CoA to yield acetyl-CoA and vanillyl-CoA, catalyzed by a *β*-ketothiolase (Figure 6F).

A novel CoA-dependent, non-*β*-oxidative pathway for ferulic acid cleavage has also been identified (Gasson et al., 1998; Narbad and Gasson, 1998; Overhage et al., 1999). In this case, ferulic acid is activated to the CoA thioseter, catalyzed by feruloyl-CoA synthetase. Feruloyl-CoA is subsequently hydrated and non-oxidatively cleaved to vanillin and acetyl-CoA. Both reactions are catalyzed by one distinct enzyme, designated as 4-hydroxycinnamoyl-CoA hydratase/lyase (HCHL) (Mitra et al., 1999) (Figure 6G).

Process Patent Applications for the Microbial Production of Natural Vanilla from Ferulic Acid

Gross et al. (1991) patented the bioconversion of ferulic acid to vanillin by *Pycnoporus cinnabarinus* CNCM n DEG I-937 or CNCM n DEG I-938 in which a maximum conversion yield of 20.5% was observed. The same process was also patented in USA (Gross et al. 1993). In another patent, *Pseudomonas putida* ATCC55180 resulted in the production of vanillin upon ferulic acid biotransformation, in the presence of sulfhydryl compounds (Labuda et al. 1992). A process combining the biotransformation of sugar-beet-pulp ferulic acid to vanillic acid by *Aspergillus niger* and the biotransformation of recovered vanillic acid to vanillin by *P. cinnabarinus* led to a patent by Lesage-Meessen et al. (1994). Audras and More, (1996) patented a process for the bioconversion of ferulic acid to vanillin and vanillic acid by *Streptomyces setonii* (strain CNCM n I-1555) immobilized in calcium alginate beads. Vanillin with a molar yield of 53-69% was obtained within 72 h of incubation. An extensive patent was disclosed by Narbad et al. (1997) from the Institute of Food Research. They described the enzymes responsible for the conversion of ferulic acid to vanillin and vanillic acid in *Pseudomonas fluorescens* AN103. Müller et al. (1998), disclosed a process for the bioconversion of ferulic acid to vanillin by *S. setonii* strain ATCC 39116. Vanillin with a molar yield of 75% was obtained after 17 h of incubation. Rabenhorst and Hopp, (1997) described the isolation of a new *Amycolatopsis* sp. strain HR167 (DSM9991 or DSM9992) that was able to convert ferulic acid to vanillin with a molar yield of 77.8% within 32 h of incubation. In 1998, the cloning and sequencing of a ferulic acid decarboxylase gene from *Saccharomyces cerevisiae* W3 was described by Ago and Kikuchi (1998). Cheetham et al. (2000) with Zylepsis Ltd., described the release of ferulic acid from *Rhizopus oryzae* by ferulic acid esterase and the subsequent conversion of ferulic acid to vanillic acid, vanillyl alcohol, and vanillin by several microorganisms. Sun et al. (2008) was granted a patent by the USA for the method producing vanillic acid and vanillin from waste residue of rice bran oil by fermentation and biotransformation. *Aspergillus niger* CGMCC 0774 is used to hydrolyze waste residue oil to produce ferulic acid and further transform the ferulic acid into vanillic acid. Then *Pycnoporus cinnabarinus* CGMCC 1115 is added into the vanillic acid-enriched culture media to perform a reductive conversion reaction, resulting in the production of vanillin. The same process was also patented in Europe (Sun et al. 2009).

Conclusion

Various methods employed for biotechnological production of natural vanilla metabolites upon ferulic acid biotransformation by plant tissue culture and microorganisms have been discussed. In the case of plant tissue culture-mediated transformation, although production of vanilla flavor metabolites was demonstrated in all cases, the low yields and the relatively high costs led to commercial infeasibility. Plant cell and organ culture techniques demand stringent attention to sterility and the growth rates are much slower than in microbial cultures.

Microorganisms can biotransform ferulic acid into vanillate derivatives by various chain-shortening routes as discussed above. In most of the microorganisms studied so far, vanillin was further oxidized to vanillic acid, which either underwent *O*-demethylation leading to the formation of protocatechuic acid or decarboxylation to guaiacol. This reflects the problem of detecting a stable and substantial amount of vanillin and other metabolites in the culture medium. Thus, to obtain a nature-isolated microorganism capable of producing stable components of natural vanilla, is always a desired aim to be fulfilled.

REFERENCES

Achterholt S, Priefert H, Steinbuchel A. Identification of *Amycolatopsis* sp. Strain HR167 genes, involved in the bioconversion of ferulic acid to vanillin. *Appl Microbiol Biotechnol* 2000;54:799-807.

Adamu HA, Iqbal S, Chan KW, Ismail M. Biotransformation of ferulic acid to 4-vinyl guaiacol by *Lactobacillus farciminis. Afr J Biotechnol* 2012;11:1177-84.

Ago S, Kikuchi Y. (1998) *Ferulic acid decarboxylase.* Patent application US5955137.

Ander P, Hatakka A, Eriksson KE. Vanillic acid metabolism by the white-rot fungus *Sporotrichum pulverulentum. Arch Microbiol* 1980;125:189-202.

Ander P, Mishra C, Farrell RL, Eriksson KE. Redox reactions in lignin degradation: interactions between laccase, different peroxidases and cellobiose:quinone oxidoreductase. *J Biotechnol* 1990;13:189-98.

Andreoni V, Bestetti G. Ferulic acid degradation encoded by a catabolic plasmid. *FEMS Microbiol Ecol* 1988;53:129-32.

Andreoni V, Galli E, Galliani G. Metabolism of ferulic acid by a facultatively anaerobic strain of *Pseudomonas cepacia. Syst Appl Microbiol* 1984;5:299-304.

Audras B, More J. (1996) *Procédé de production de vanilline à partir de précurseurs benzéniques par bioconversion.* Patent application WO9634971.

Baqueiro-Peña I, Rodríguez-Serrano G, González-Zamora E, Augur C, Loera O, Saucedo-Castañeda G. Biotransformation of ferulic acid to 4-vinylguaiacol by a wild and a diploid strain of *Aspergillus niger. Biores Technol* 2010;101:4721-24.

Barghini P, Montebove F, Ruzzi M, Schiesser A. Optimal conditions for bioconversion of ferulic acid into vanillic acid by *Pseudomonas fluorescens* BF13 cells. *Appl Microbiol Biotechnol* 1998;49:309-14.

Barthelmebs L, Divies C, Cavin J-F. Knockout of the *p*-coumarate decarboxylase gene from *Lactobacillus plantarum* reveals the existence of two other inducible enzymatic activities involved in phenolic acid metabolism. *Appl Environ Microbiol* 2000;66:3368-75.

Beilen JBV, Li Z. Enzymes Technology: an overview. *Curr Opin Microbiol* 2002;13:334-8.

Bonnin E, Lesage-Meessen L, Asther M, Thibault J-F. Enhanced bioconversion of vanillic acid into vanillin by the use of "natural" cellobiose. *J Sci Food Agric* 1999;79:484-86.

Brunati M, Marinelli F, Bertolini C, Gandolfi R, Daffonchio D, Molinari F. Biotransformation of cinnamic and ferulic acid with actinomycetes. *Enzyme Microbial Technol* 2004;34:3-9.

Cavin J-F, Dartois V, Diviès C. Gene cloning, transcriptional analysis, purification and characterization of phenolic acid decarboxylase from *Bacillus subtilis*. *Appl Environ Microbiol* 1998;64:1466-71.

Cerrutti P, Alzamora SM, Vidales SL. Vanillin as an antimicrobial for producing shelf-stable strawberry puree. *J Food Sci* 1997;62:608-10.

Cheetham P. The use of biotransformation for the production of flavours and fragrances. *Trends Biotechnol* 1993;11:478-88.

Cheetham PSJ, Gradley ML, Sime JT. (2000) *Flavour/aroma materials and their preparation.* Patent application WO0050622.

Chow KT, Pope MK, Davies J. Characterization of a vanillic acid non-oxidative decarboxylation gene cluster from *Streptomyces* sp. D7. *Microbiology* 1999;145:2393-2403.

Civolani C, Barghini P, Roncetti AR, Ruzzi M, Schiesser A. Bioconversion of ferulic acid into vanillic acid by means of a vanillate-negative mutant of *Pseudomonas fluorescens* strain BF13. *Appl Environ Microbiol* 2000;66:2311-17.

Clark GS. Vanillin. *Perfumer Flavorist* 1990;15:45-54.

Degrassi G, DeLaureto PP, Bruschi CV. Purification and characterization of ferulate and *p*-coumarate decarboxylase from *Bacillus pumilus*. *Appl Environ Microbiol* 1995;61:326-32.

Delneri D, Degrassi R, Rizzo R, Bruschi CV. Degradation of *trans*-ferulic acid and *p*-coumaric acid by *Acinetobacter calcoaceticus* DSM 586. *Biochem Biophys Acta* 1995;1244:363-67.

Dignum MJW, Kerler J, Verpoorte R. Vanilla production: technological, chemical, and biosynthetic aspects. *Food Res Inter* 2001;17:199-219.

Falconnier B, Lapierre C, Lesage-Meessen L, Yonnet G, Brunerie P, Colonna Ceccaldi B, et al. Vanillin as a product of ferulic acid bioconversion by the white-rot fungus *Pycnoporus cinnabarinus* 1-937 identification of metabolic pathways. *J Biotechnol* 1994;37:123-132.

Funk C, Brodelius P. Phenylpropanoid metabolism in suspension cultures of *Vanilla planifolia* Andr. II. Effects of precursor feeding and metabolic inhibitors. *Plant Physiol* 1990;94:95-101.

Funk C, Brodelius P. Vanilla planifolia Andrews: *in vitro* biosynthesis of vanillin and other phenylpropanoid derivatives. In Bajaj YPS, (Ed). *Biotechnology in Agriculture and Forestry.* Medicinal and aromatic plants VI, Springer, Berlin, 1994:377-402.

Gasson MJ, Kitamura Y, McLauchian WR, Narbad A, Parr AJ, Parsons EL, et al. Metabolism of ferulic acid to vanillin. A bacterial gene of the enoyl-SCoA hydratase/isomerase superfamily encodes an enzyme for the hydration and clease of a hydroxycinnamic acid SCoA thioster. *J Biol Chem* 1998;273:4163-70.

Ghosh S, Sachan A, Mitra A. Degradation of ferulic acid by a white-rot fungus *Schizophyllum commune. World J Microbiol Biotechnol* 2005;21:385-8.

Ghosh S, Sachan A, Mitra A. Formation of vanillic acid from ferulic acid by *Paecilomyces variotii* MTCC 6581. *Curr Sci* 2006;90:825-9.

Ghosh S, Sachan A, Sen SK, Mitra A. Microbial transformation of ferulic acid to vanillic acid by *Streptomyces sannanensis* MTCC 6637. *J Ind Microbiol Biotechnol* 2007;34:131-8.

Grbic-Galic D. Fermentative and oxidative transformation of ferulate by a facultatively anaerobic bacterium isolated from sewage sludge. *Appl Environ Microbiol* 1985;50:1052-7.

Gross B, Asther M, Corrieu G, Brunerie P. (1991) *Production of vanillin by bioconversion of benzenic precursors.* Patent application EP0453368.

Gross B, Asther M, Corrieu G, Brunerie P. (1993) *Production of vanillin by bioconversion of benzenoid precursors by Pycnoporus.* Patent application US5262315.

Gupta JK, Hamp SG, Buswell JA, Eriksson KE. Metabolism of trans-ferulic acid by the white-rot fungus *Sporotrichum pulverulentum. Arch Microbiol* 1981;128:349-54.

Gurujeyalakshmi G, Mahadevan A. Dissimilation of ferulic acid by *Bacillus subtilis. Curr Microbiol* 1987;16:69-73.

Havkin-Frenkel D, Podstolski A, Knorr D. Effect of light on vanillin precursors formation by *in vitro* cultures of *Vanilla planifolia. Plant Cell Tiss Org* 1996;45:133-6.

Havkin-Frenkel D, Podstolski A, Witkowska E, Molecki P, Mikolajczyk P. Vanillin biosynthetic pathways, an overview. In Fu TJ, Singh G, Curtis WR, (Eds). *Plant Cell and Tissue Culture for the Production of Food Ingredients,* Kluwer Academic Press/ Plenum Publishers, New York, 1999:35-43.

Huang Z, Dostal L, Rosazza JPN. Microbial transformations of ferulic acid by *Saccharomyces cerevisiae* and *Pseudomonas fluorescens. Appl Environ Microbiol* 1993;59:2244-50.

Huang Z, Dostal L, Rosazza JPN. Mechanism of ferulic acid conversion to vanillic acid and guaiacol by *Rhodotorula rubra*. *J Biol Chem* 1994;268:23594-98.

Johnson TS, Ravishankar GA, Venkataraman LV. Biotransformation of ferulic acid and vanillylamine to capsaicin and vanillin in immobilized cell cultures of *Capsicum frutescens*. *Plant Cell Tiss Org* 1996;44:117-21.

Karmakar B, Vohra RM, Nandanwar H, Sharma P, Gupta KG, Sobti RC. Rapid degradation of ferulic acid via 4-vinylguaiacol and vanillin by a newly isolated strain of *Bacillus coagulans*. *J Biotechnol* 2000;80:195-202.

Krings U, Pilawa S, Theobald C, Berger RG. Phenyl propenoic side chain degradation of ferulic acid by *Pycnoporus cinnabarinus*- elucidation of metabolic pathways using [5-^{2}H]-ferulic acid. *J Biotechnol* 2001;85:305-14.

Labuda IM, Goers SK, Keon KA. (1992) *Bioconversion process for the production of vanillin*. Patent application US 5128253.

Labuda IM, Keon KA, Goers SK. Microbial bioconversion process for the production of vanillin. In Schreier P, Winterhalter P, (Eds). *Progress in flavour and precursor studies,* Allured Publishing, Carol Stream, IL, 1993: 477-482.

Lee IY, Volm TG, Rosazza JPN. Decarboxylation of ferulic acid to 4-vinylguaiacol by *Bacillus pumilus* in aqueous-organic solvent two-phase systems. *Enzyme Microb Technol* 1998;23:261-66.

Lesage-Meessen L, Delattre M, Haon M, Asther M. (1994) *Procédé d'obtention d'acide vanillique et de vanilline par bioconversion de microorganismes filamenteux*. Patent application WO 9608576.

Lesage-Meessen L, Delattre M, Haon M, Thibault JF, Colonna Ceccaldi B, Brunerie P, et al. A two-step bioconversion process for vanillin production from ferulic acid combining *Aspergillus niger* and *Pycnoporus cinnabarinus*. *J Biotechnol* 1996;50:107-13.

Lesage-Meessen L, Haon M, Delattre M, Thibault J-F, Colonna Ceccaldi B, Asther M. An attempt to channel the biotransformation of vanillic acid into vanillin by controlling methoxyhydroquinone formation in *Pycnoporus cinnabarinus* with cellobiose. *Appl Microbiol Biotechnol* 1997;47:393-397.

Li X, Yang J, Li X, Gu W, Huang J, Zhang KQ. The metabolism of ferulic acid via 4-vinylguaiacol to vanillin by *Enterobacter* sp. Px6-4 isolated from Vanilla root. *Process Biochem* 2008;43:1132-7.

Lomascolo A, Stentelaire C, Asther M, Lesage-Meessen L. Basidiomycetes as new biotechnological tools to generate natural aromatic flavours for the food industry. *Trends Biotechnol* 1999;17:282-9.

Masai E, Harada K, Peng X, Kitayama H, Katayama Y, Masao M. Cloning and characterization of ferulic acid catabolic genes of *Sphingomonas paucimobilis* SYK-6. *Appl Environ Microbiol* 2002;68:4416-24.

Mitra A, Kitamura MJ, Gasson A, Narbad A, Parr J, Payne J, et al. 4-Hydroxycinnamoyl-Coa hydratase/lyase (HCHL)-an enzyme of phenylpropanoid chain cleavage from *Pseudomonas. Arch Biochem Biophys* 1999;365:10-6.

Muheim A, Lerch K. *Towards a high-yield bioconversion of ferulic acid to vanillin. Appl Microbiol Biotechnol* 1999;51:456-61.

Müller B, Münch T, Muheim A, Wetli M. (1998) *Process for the production of vanillin.* Patent application EP0885968.

Narbad A, Gasson MJ. Metabolism of ferulic acid via vanillin using a novel CoA-dependent pathway in a newly-isolated strain of *Pseudomonas fluorescens. Microbiology* 1998;144:1397-1405.

Narbad A, Rhodes MJC, Gasson MJ, Walton NJ. (1997) *Production of vanillin.* Patent application WO9735999.

Nishida A, Fukuzumi T. Formation of coniferyl alcohol from ferulic acid by the white-rot fungus *Trametes. Phytochemistry* 1978;17:417-9.

Oddou J, Stentelaire C, Lesage-Meessen L, Asther M, Colonna Ceccaldi B. Improvement of ferulic acid bioconversion into vanillin by use of high-density cultures of *Pycnoporus cinnabarinus. Appl Microbiol Biotechnol* 1999;53:1-6.

Ohmiya K, Takeuchi M, Chen W, Shimizu S, Kawakami H. Anaerobic reduction of ferulic acid to dihydroferulic acid by *Wolinella succinogenes* from cow rumen. *Appl Microbiol Biotechnol* 1986;23:274-9.

Ötuk G. Degradation of ferulic acid by *Escherichia coli. J Ferment Technol* 1985;63:501-6.

Ou S, Kwok KC. Ferulic acid: pharmaceutical functions, preparation and application in food. *J Sci Food Agric* 2004;84:1261-9.

Overhage J, Priefert H, Steinbuchel A. Biochemical and genetic analyses of ferulic acid catabolism in *Pseudomonas* sp. strain HR199. *Appl Environ Microbiol* 1999;65:4837-47.

Peng X, Misawa N, Harayama S. Isolation and characterization of thermophillic bacilli degrading cinnamic, 4-coumaric, and ferulic acids. *Appl Environ Microbiol* 2003;69:1417-27.

Plaggenborg R, Steinbuchel A, Priefert H. The coenzyme A-dependent, non-beta-oxidation pathway and not direct deacetylation is the major route for ferulic acid degradation in *Delftia acidovorans. FEMS Microbiol Lett* 2001;205:9-16.

Pometto AL, Crawford DL. Whole-Cell bioconversion of vanillin to vanillic acid by *Streptomyces viridosporus. Appl Environ Microbiol* 1983;45:1582-5.

Priefert H, Rabenhorst J, Steinbüchel A. Biotechnological production of vanillin. *Appl Microbiol Biotechnol* 2001;56:296-314.

Rabenhorst J, Hopp R. (1997) *Process for the preparation of vanillin and suitable microorganisms.* Patent application EP0761817.

Rahouti M, Seigle-Murandi F, Steiman R, Eriksson KE. Metabolism of ferulic acid by *Paecilomyces variotii* and *Pestalotia palmarum*. *Appl Environ Microbiol* 1989;55:2391-8.

Ramachandra Rao S, Ravishankar GA. Biotransformation of isoeugenol to vanilla flavour metabolites and capsaicin in suspended and immobilized cell cultures of *Capsicum frutescens*: study of the influence of beta-cyclodextrin and fungal elicitor. *Process Biochem* 1999;35:341-8.

Ramachandra Rao S, Ravishankar GA. Vanilla flavour: production by conventional and biotechnological routes. *J Sci Food Agric* 2000;80:289-304.

Ranadive AS. Vanilla-cultivation, curing, chemistry, technology and commercial products. In Charalambous G. (Ed.), *Spices, Herbs and Edible Fungi,* Elsevier, Amsterdam, 1994:517-577.

Romagnoli LG, Knorr D. Effects of ferulic acid treatment on growth and flavour development of cultured *Vanilla planifolia* cells. *Food Biotechnol* 1988;2:93-104.

Rosazza JPN, Huang Z, Dostal L, Volm T, Rousseau B. Review: Biocatalytic transformations of ferulic acid: an abundant aromatic natural product. *J Ind Microbiol* 1995;15:457-71.

Stentelaire C, Lesage-Meessen L, Oddou J, Bernard O, Bastin G, Colonna Ceccaldi B, et al. Design of a fungal bioprocess for vanillin production from vanillic acid at scalable level by *Pycnoporus cinnabarinus*. *J Biosci Bioeng* 2000;89:223-30.

Sun R, Sun XF, Wang SQ, Zhu W, Wang XY. Ester and ether linkages between hydroxycinnamic acids and lignin from wheat, rice, rye and barley straws, maize stems and fast-growing poplar wood. *Ind Crops Prod* 2002;15:179-88.

Sun Z, Zheng P, Guo X, Lin G, Yin H, Wang J, Bai Y (2008). *Method for producing vanillic acid and vanillin from waste residue of rice bran oil by fermentation and biotransformation.* Patent application US7462470.

Sun Z, Zheng P, Guo X, Lin G, Yin H, Wang J, Bai Y (2009). *Method for producing vanillic acid and vanillin from waste residue of rice bran oil by fermentation and biotransformation.* Patent application EP1734128.

Suresh B, Ritu T, Ravishankar GA. Vanilla flavour production through biotransformation using *Capsicum frutescens* root cultures. *Biocatal Biotransform* 2003;21:333-40.

Sutherland JB, Crawford DL, Pometto AL. Catabolism of cinnamic, *p*-coumaric acid and ferulic acids by *Streptomyces setonii*. *Canadian J Microbiol* 1983;29:1253-7.

Tillett R, Walker JRL. Metabolism of ferulic acid by a *Penicillium sp. Arch Microbiol* 1990;154:206-8.

Tome A, Wood JM. The degradation of *trans*-ferulic acid by *Pseudomonas acidovorans*. *Biochemistry* 1970;9:337-43.

Topakas E, Kalogeris E, Kekos D, Macris BJ, Christakopoulos P. Bioconversion of ferulic acid into vanillic acid by the thermophilic fungus *Sporotrichum thermophile. Lebensmittel-Wissenschaft und-Technologie* 2003;36:561-5.

Tripathi U, Ramachandra RS, Ravishankar GA. Biotransformation of phenylpropanoid compounds to vanilla flavor metabolites in cultures of *Haematococcus pluvialis*. *Process Biochem* 2002;38:419-26.

Tsujiyama S, Ueno M. Formation of 4-vinyl guaiacol as an intermediate in bioconversion of ferulic acid by *Schizophyllum commune*. *Biosci Biotechnol Biochem* 2008;72: 212-5.

Walton NJ, Mayer MJ, Narbad A. Vanillin. *Phytochemistry* 2003;63:505-15.

Westcott RJ, Cheetham PSJ, Barraclough AJ. Use of organized viable Vanilla plant aerial roots for the production of vanillin. *Phytochemistry* 1994;35:135-8.

In: Recent Advances in Microbiology. Volume 3 ISBN: 978-1-53614-057-6
Editors: N.C. Gautam, S.P. Tiwari and R. Sharma

Chapter 4

SURVIVAL OF PLANT GROWTH PROMOTING RHIZOBACTERIA (PGPR)

Anurag Yadav[1,*] and Kususm Yadav[2]
[1]College of Basic Science and Humanities,
S.D. Agricultural University, Gujarat, India
[2]Department of Biochemistry, University of Lucknow,
Lucknow, Uttar Pradesh, India

INTRODUCTION

The Plant Growth Promoting Rhizobacteria (PGPR) are bacterial species that live in the rhizosphere (root zone) of a plant and increase plant growth. This group includes all the bacteria that help plant growth either directly or indirectly, through various means. The term PGPR was used to describe the bacteria that colonize plant roots following inoculation on seeds to enhance plant growth (Kloepper and Schroth, 1978). Plants interact with an enormous volume of microorganisms by way of their root system, due to the richness of organic matter in the soil. Having existed in the proximity of roots since times immemorial, microorganisms have evolved various strategies to interact with plants for availing nutrients. Some microorganisms interact by colonizing the root surface to avail root exudates (Bais et al., 2006). Others interact strongly by secreting a spectrum of metabolites and can modify the rhizosphere properties. These changes, in turn, affect plant growth directly or indirectly (Lynch, 1990). Possible ways by which the microbiota can interact with plants are by (i) providing nutrients to plants, (ii) growth stimulation of plants by secretion of growth hormones, (iii) controlling plant pathogens,

[*]Corresponding Author EMail: yadav123@gmail.com.

(iv) improving soil quality, (v) minimizing pollutants through bioremediation and by (v) bioaccumulation or microbial leaching of inorganic compounds. It is generally considered that rhizosphere is dominated by non-fermenting, Gram-negativerods, especially pseudomonads.

The phenomenon of plant growth promotion is governed by a variety of bacterial traits and specific genes, which include chemotaxis toward seed and root exudates, motility, production of pili or fimbriae (Lugtenberg et al., 2001), production of specific cell surface components, ability to use specific components of root exudates, protein secretion and quorum sensing (Liu et al., 2007). Latest screening methods based on gene fusion technologies are aiding the identification of new, previously uncharacterized genes. Altering the mutants for expression of traits is helping in understanding the exact role of each gene (Lugtenberg et al., 2001). These methods use reporter transposons (Roberts et al., 1998) and *in vitro* expression technology (IVET) (Rainey, 1999) to detect genes expressed during colonization.

Inoculant survival in the rhizosphere is a prerequisite for successful plant growth promotion. Often the soil inoculants that perform well in laboratory fail to produce satisfactory results in the field (Capper and Campbell, 1986). These challenges restrict successful commercialization of PGPR (Mathre et al., 1999) as most of the available agro-industrial rhizobacterial strains face population decline after their introduction in soil (Howie and Echandi, 1983; Harvey Bolton et al., 1991). Also, various soil types, ranging from fertile to non-fertile ones exist across the globe; therefore, the strain successful in one place may not perform equally well in another patch of land. The approach of selectively screening advantageous native soil microflora and its mass scale reintroduction in the rhizosphere can aid in enhancing bioinoculant survival period (Young et al., 1995).

A wide range of bacteria was successfully tested to increase plant growth. Some bacterial genera like *Alcaligenes*, *Arthrobacter*, *Azospirillum*, *Azotobacter*, *Bacillus, Burkholderia*, *Enterobacter*, *Klebsiella*, *Pseudomonas* and *Serratia*are well known to enhance plant growth. Their commercial formulations as bioinoculants are available for use on agricultural lands. However, the majority of literature has focused on *Pseudomonas* and *Bacillus*. The *Pseudomonas* genus has evolved as a promising rhizobacteria, which enhances plant growth by several mechanisms, enabling them to utilize various substrates released by the roots. Microorganisms like *Pseudomonas putida* are competitively more stable in rhizosphere because they secrete heterologous siderophores (Raaijmakers et al., 1995). Often such inoculants perform well under nutrient-deficient conditions. Carbon-starved *Pseudomonas fluorescens* R2f populations have been reported to enhance resistance to otherwise lethal treatments, like exposure to ethanol, high temperature, osmotic tension and oxidative stress (van Overbeek et al., 1995). Most of such non-sporing PGPR require special carriers for survival at the time for their mass cultivation. However, working with spore-forming populations is more convenient due to their higher survival rate and easier mass preparation. The genus *Bacillus*, an endospore producer, is tolerant to desiccation and heat. It has a longer life in commercial formulations and is a genus of choice in the biofertilizer industry. Bacterial spore-formers have short generation time, greater mobility and the capability to colonize roots. Moreover, they produce a wide variety of secondary metabolites including plant growth-regulating substances and antibiotic compounds (Kloepper, 1994).

FACTORS AFFECTING BACTERIAL SURVIVAL IN SOIL

The population of rhizobacteria like other microorganisms is affected by factors like varied nutritional capabilities of soil, varied atmospheric conditions, antagonism by native rhizosphere microflora, competition for soil nutrition with native microflora and mutations in PGPR strains to become non-PGPRs. In rhizosphere, microorganisms interact mostly by combined niche exploitation rather than by direct competition (Goddard et al., 2001). Root and vesicular-arbuscular mycorrhizal fungal component enhance water-stable soil aggregates (WSA), which in turn, indirectly affect the number of microorganisms by providing favorable and practical habitat through the creation of habitable pore space in WSA (Andrade et al., 1998). Table 1 shows the extensive list of factors that influence the rhizobacterial survival in soil. Generally, PGPR population sharply declines after getting introduced in the rhizosphere (Compeau et al., 1988). In an experiment on *Azospirillumbrasilense* Cd and Sp-245 introduced in the rhizosphere of wheat and tomato plants, the inoculant population declined sharply and reached below detection limit when plants were removed from the site (Bashan et al., 1995). Compared to field soil, the inoculant population decreases faster in laboratory soil (Thompson et al., 1992), and, therefore a record of observation site is important for a bioinoculant population study. In a number of reports, the introduced bacterial population reached below the detection limit (Figure 1) (Lucas and Davis, 1961; Frey-Klett et al., 1997; Bais et al., 2006). This decline could be the result of physical stress (Chen and Alexander, 1973; Wessendorf and Lingens, 1989), chemical stress (Acea et al., 1988) or biological stress (Acea et al., 1988; Heijnen and van Veen, 1991), which are encountered by the introduced species in soil. Our current knowledge on bioinoculant survival is limited and mainly based on biotic and abiotic factors of soil. Biological factors are the main cause of inoculant population decline, whereas abiotic factors like soil texture type, pH, temperature and moisture (Dunn et al., 1985) directly affect rhizobacteria by imparting a variety of stresses (Gu and Mazzola, 2001). Aforesaid factors indirectly affect the activity of indigenous microflora. No population decline has been reported in sterilized soil (Heynen et al., 1988), which proves that biotic factors play an important role in shaping the microbial ecology.

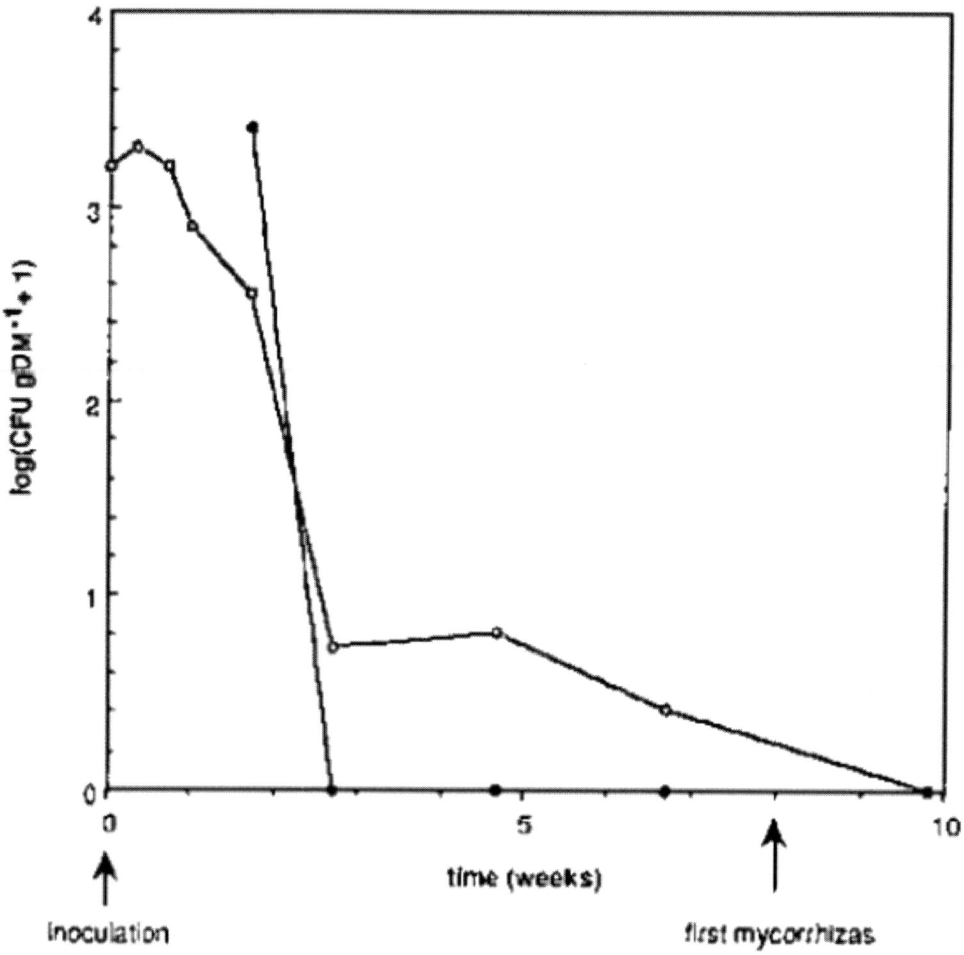

Figure 1. Population dynamics of *Pseudomonas fluorescens* BBc6R8 in the soil (open circles) and in the rhizosphere (closed circles) of Douglas fir seedlings-*Laccaria bicolor* mycorrhizas in the nursery experiment. Reprinted from reference (Frey-Klett et al., 1997) with permission.

The scarcity of available nutrients and environmental variations also lead to bioinoculant population decline. The ineffectiveness of PGPR in the field has often been attributed to its inability to colonize plant roots. (Benizri et al., 2001; Bloemberg & Lugtenberg, 2001). In one of the unique studies in this regard, the mathematical modeling of inoculant was done to simulate the growth of PGPR in the rhizosphere. The results of simulation indicated that in the rhizosphere, competition for limited resources between the introduced population and the resident microorganisms was the most important factor in determining PGPR survival (Strigul and Kravchenko, 2006). Competition between inoculant and native microflora severely affects the indigenous microbial population. Slow-growing bioinoculants are at a disadvantage in the competition-filled micro-ecosystem. They are heavily grazed upon by protozoal and nematode populations, and hence, may not get sufficient time to multiply.

Table 1. List of factors affecting rhizobacterial survival in soil

Factors	Name of the factor	Mode of action	Reference
Chemical factors	Soil pH	Influence the nutrient availability in soil	(Johri et al., 2003; Tyler and Triplett, 2008)
	Oxygen supply	Influence the redox reactions of site	(Sylvia et al., 2004)
	Salinity	Suppress the root tip colonization of microorganisms[1], accumulation of sodium and pigments and reduction of antioxidants and potassium[2]	[1](Gupta Sood, 2003)· [2](Anburaj et al., 2012)
	Cation exchange capacity	Influence the nutrient availability of soil	(Rhoades, 1982)
	Carbon starvation	Induce tolerance to variety of environmental stress[1], enhance the high-temperature tolerance of microorganisms[2]	[1](Gu and Mazzola, 2001)· [2](van Overbeek et al., 1995)
	Bioavailability of substrates	Sucrose availability stimulates high levels of uptake and accumulation of glycine betaine (betaine) in *Rhizobium*	(Kayser et al., 2001)
Physical factors	Soil texture	Related to oxygen diffusion property	(Bashan et al., 1995; Tyler and Triplett, 2008)
	Soil pore space	Vital role in water and air movement	(Bashan et al., 1995; Or et al., 2007)
	Soil water/moisture	Establishment of microbial community	(Gupta Sood, 2003)
	Soil temperature	*Rhizobacterial chemotaxis* towards roots	(Gupta Sood, 2003)
Biological factors	Protozoa in soil	Predation of bacteria	(Heynen et al., 1988; England et al., 1993; Griffiths, 1994)
	Nematodes in soil	Predation of bacteria	(Griffiths, 1994)
	Earthworm activity in soil	Influence metabolic activity of microorganisms by secreting nitrogen in soil	(Doube et al., 1994; Jusselme et al., 2012)
	Plant-Microbe interactions in soil	Increase plant growth and health which positively affects rhizobacterial survival	(Kuiper et al., 2004; Berg, 2009)
	Microbial genetic modification	Duplication of catabolic genes of *Arthrobacter aurescens* strain TC1 increases its metabolic ability	(Mongodin et al., 2006)
	Presence of ACC deaminase in bacteria	Facilitate plant growth to overcome abiotic stress	(Shakir et al., 2011)
	Bacterial operons	The *phaAB*$\rightarrow$*G* operon of *Azospirillum brasilense* CBG497 is involved in its adaptation to alkaline pH in the presence of K^+	(Acosta-Cruz et al., 2012)
	Presence of specific chemotactic traits by bacteria	Cause movement ofbacteria towards root exudates and seeds	(Soby and Bergman, 1983; Lugtenberg et al., 2001)
	Presence of bacterial flagella	Increased movement of bacteria increase chance of survival	(Soby and Bergman, 1983; Boelens et al., 1994; Turnbull et al., 2001)
	Presence of bacterial pili or fimbriae	Improve bacterial adherence to plant surface	(Bitton and Marshall, 1980; Lugtenberg et al., 2001)
	Low-temperature tolerance by bacteria	Production of antifreeze protein by *Pseudomonas putida* GR12-2	(Sun et al., 1995)

The proper selection of physiological traits is important for bioinoculant survival and their effective root colonization. Majority of survival studies have been conducted on non-spore-forming genera like *Pseudomonas* (Dandurand et al., 1994; de Leij et al., 1994; Frey-Klett et al., 1997; Perotti et al., 2005; Abat, 2006; Bais et al., 2006), *Azospirillum* (Perotti et al., 1985; Eory et al., 1995; Gonzalez, 1999; Joe et al., 2012) or *Rhizobium* (Baldani and Weaver, 1992) and mostly on single plant species. In an exclusive study on 690 fluorescent pseudomonads, it was found that colonization ability of some transient strains increased appreciably when inoculated with rhizosphere-competent isolates (Goddard et al., 2001).

Rhizosphere not only supports PGPR and other commensal bacteria but some human pathogens also survive there (Williams et al., 2007). Genera closely related to *Salmonella* and *Escherichia* (e.g., *Klebsiella* and *Serratia*) contain species, which are pathogenic to humans and are routinely isolatable from plant tissues and the rhizosphere (Tyler and Triplett, 2008).

Chemical Factors

pH is known to influence the availability of nutrients in the soil (Lucas and Davis, 1961), playing an immense role in shaping the soil microflora. At specific pH, the chemical specification of soil determines the availability of nutrients. It fluctuates due to biotic and abiotic processes of soil. Consumption or release of H^+ by a redox reaction, fermentation and rainfall leaches bases in the soil, which lowers the soil pH (Tyler and Triplett, 2008). Change of pH can hinder the plant growth-promoting traits of rhizobacteria by interfering with their metabolics. However, some bacteria like *Halomonas* and *Archaeoglobus* can tolerate extremes of pH. Change of pH has been reported to affect the accessibility of metals and organic matter to microorganisms (Tyler and Triplett, 2008). Slightly acidic to neutral soil pH (6, 6.5 and 7) favors the chemotaxis of PGPR towards roots (Gupta Sood, 2003). Root exudates are also known to alter the nutrient availability of soil (Bais et al., 2006).

Cation exchange capacity (CEC) is another factor that directly affects nutrient availability in soil by influencing charged nutrients. CEC represents a number of exchangeable cations that soil can hold at specific pH. Negatively charged soil particles can attract or repel soil microorganisms and can, thus, influence nutrient availability. The process of CEC is not fully understood yet, but some of the proposed mechanisms include ion exchange attractions, weak attractive forces, coordinated bonding and hydrogen bonding.

Several nutrients and micronutrients in soil regulate bioinoculant survival. In an experiment on heavily nitrogen-fertilized maize crop with phytostimulator *Azospirillum lipoferum* CRT1, no significant plant growth enhancement was observed; nevertheless, the inoculant survived well with a final population recovery of 6 log CFU g^{-1} root from an initial population of 8 log CFU g^{-1}, after 142 days of inoculation (El Zemrany et al., 2006). Sucrose has also been shown to enhance survival of certain *Rhizobium* spp. and stimulate high-level glycine betaine uptake and accumulation (Mugnier and Jung, 1985). Bacterial survival and proliferation in the environment as well as within various hosts are critically dependent on the uptake and sequestration of transition metals like manganese, zinc, and iron (Jing et al., 2007). Bacterial cells can stringently regulate intracellular zinc levels, but the higher concentrations are toxic to cellular functions. Rhizobacteria cope with toxic zinc concentrations by transporting it outside the cell. These bacteria have evolved protein types that are involved in binding and transport of zinc (Claverys, 2001).

PHYSICAL FACTORS

Based on particle size, the soil can be classified as sand (0.05-2.0 mm), silt (0.002 – 0.05mm) and clay (< 0.002 mm) (Tyler and Triplett, 2008). Percentages of clay, nitrogen and organic matter shape the soil texture. Moreover, water-holding capacity of the soil can be positively correlated with bacterial viability (Bashan et al., 1995). Observations have shown the influence of soil type on the microbial community (Vandenhove et al., 1991; Höflich et al., 1995; Zachow et al., 2011; Dey et al., 2012). Clay, the finest soil, holds rich organic matter content, but has low oxygen diffusion property. Sandy soil, on the other hand, has good oxygen diffusion but has the lowest water holding capacity. Due to the differences in soil composites and particle size, the bioinoculant survival rate varies. Observation has shown higher inoculant survival rate in clayey soil than in sandy soil (Bashan, 1986). This is because soil type affects catabolic gene expression of inoculated bacterial strains (Afzal et al., 2011).

The soil pore size is relatable to its particle size and plays a vital role in water and air movement. The soil pore space is majorly determined by size and arrangement of aggregates, which in turn, affects movement of water, air and microorganisms in the soil. Soil pores can be divided into five types *viz.*, macropores (> 75μm), mesopores (> 30-70μm), micropores (5-30 μm), ultra-micropores (0.1-5 μm) and crytopores (< 0.1μm). Soil pores provide a protectivemicro-habitat for microorganisms. The primary effect of soil pore geometry and hydration status is on diffusion pathways towards and away from stationary microbial colonies (Or et al., 2007). Soil texture and water flow in porosity directly influence the bacterial pool (Mateille et al., 2009). Soil water plays a vital role in the establishment of microbial community. The study on *Pseudomonas fluoresce* sand *Azotobacter chroococcum* revealed a positive correlation between soil water potential and bacterial chemotactic response towards roots (Gupta Sood, 2003).

Temperature is another factor that greatly impacts the rate of the biological, chemical and physical process in soil. Temperature directly affects bioinoculant concentration by altering their metabolism; it does so indirectly by influencing the properties of soil. Rhizobacterial optimal temperature limit varies widely as they belong to several genera. In the study on strains of *Pseudomonas fluorescens* and *P. putida,* it was found that their population stability was temperature-specific, which maximized at a lower temperature of 12^0C (Loper et al., 1985). Exposure to carbon-limited growth conditions has been shown to induce resistance to a variety of environmental stresses in bacteria (Gu and Mazzola, 2001). Generally, the temperature of 20-30^0C in soil favors PGPR chemotaxis towards roots (Gupta Sood, 2003). Study shows that nutrient availability regulates the temperature optima of bioinoculants. For example, a carbon-starved mutant of *Pseudomonas fluorescens* R2f was found to enhance tolerance to high temperature (van Overbeek et al., 1995).

Oxygen, which is required by all aerobic microorganisms in nature, is yet another important physical factor. Level of oxygen in a rhizosphere microhabitat dictates the intensity of redox reactions of that place. Under low level of oxygen, facultative anaerobic microorganisms or other microorganisms that can utilize sulfur and iron as a source of electrons are able to survive. Salinity as a factor reduces the stimulatory effect of PGPR by suppressing the root tip colonization of bacterial strains (Egamberdieva, 2011).

The physiological status of inoculants is important for determining the efficiency of inoculant. Methods like micro-biosensor-based reporter gene insertions are available for finding the actual stress status of inoculant in the rhizosphere. Chemical methods like substrate-induced respiration with several limitations are also useful (Bashan et al., 1995). In *Pseudomonas putida* strain 2C8, adaptation to nutrient-limited conditions by repeated culture on soil agar media was found to induce resistance to osmotic tension, oxidative stress and carbon-limited culture conditions (Gu and Mazzola, 2001). Bacterial survival and proliferation in the environment as well as within various hosts are critically dependent on the uptake and sequestration of transition metals such as manganese, zinc, and iron (Jing et al., 2007). Certain rhizobacteria having 1-aminocyclopropane-1-carboxylate (ACC) deaminase can facilitate plant growth to overcome abiotic stress (Shakir et al., 2011).

BIOLOGICAL FACTORS

Soil microorganisms are sensitive to biological factors as varieties of animals live there and influence the dynamics. These animals can be divided into macrofauna (> 2 mm, earthworms), mesofauna (0.1-0.2 mm, mites and nematodes) and microfauna (< 0.1 mm, protozoa and nematodes). These eukaryotes heavily influence the soil microbial activity. Direct grazing of bacterial bioinoculants by protozoa and nematodes is the major reason for bioinoculant population decline (Heynen et al., 1988; Griffiths, 1994). A strong correlation exists between bioinoculant population decline and protozoal growth in soil (Heynen et al., 1988; Wright et al., 1995). Reports show that these organisms predate selectively on inoculants, influencing the microbial community of soil. Grazing increases decomposition rate and frees nutrients in the soil, which in turn, causes increase in overall microbial activity. Earthworms alter the soil habitat by forming burrows in the soil, which increases gas exchange, water movement and resource location. They are also the vectors of beneficial soil bacteria that influence the population dynamics of microorganisms in soil and microbial metabolic activity (Doube et al., 1994; Jusselme et al., 2012).

Microbial interaction is another factor that affects the bioinoculant population in the soil. Mutualism between the host plant and microorganism (legumes-rhizobia and plants-mycorrhizal fungi) provides a long-term survival benefit to the partner microorganisms. Processes like acid production, toxic chemical build-up (amensalism) or fermentation in flooded soil limit the population of bioinoculant. Role of deleterious rhizobacteria (DRB) in the rhizosphere is also important in rhizosphere dynamics. Deleterious rhizobacteria are predominant saprophytes that aggressively colonize plant seeds, roots and rhizosphere to metabolize organic substances released by plant tissues (Kremer, 2006). Nutrient bioavailability is yet another factor, which designates the availability of soil substrates. Bioavailability is particularly important from environmental perspectives because of its correlation with bioremediation. The three major factors that influence bioavailability are sorption, solubility and chemical specification. Sorption defines the way substrate bonds to the soil surface, especially charged surfaces. Solubility, the capacity of the substrate to dissolve in the soil, is the leading factor for nutrient bioavailability. Chemical properties of soil include pH, salinity, CEC, the percentage of organic matter and C:N ratio (carbon to nitrogen).

In addition, genetics of rhizobacteria also plays a leading role in their survival. In a study on *Arthrobacter aurescens* TC1, the survival ability of bacteria was co-relatable with its increased metabolic ability due to duplication of catabolic genes. The bacteria *A. aurescens* was successful in funneling plasmid-born genetic metabolic intermediates to chromosomal encoded pathways (Mongodin et al., 2006). Results also suggest the involvement of plasmids in mediating the response to stress conditions and enhancing bacterial survival (Baldani and Weaver, 1992). In *Azospirillum brasilense* CBG497, the survival ability was correlated with a large 1.59 Mbp plasmid. The presence of 62 genes in the plasmid represent the minimal gene set and chromid core genes suggest their importance in bacterial survival. Further, *phaAB* $\rightarrow$ *G* operon of the bacteria was reported to be involved in the bacterial adaptation to alkaline pH in the presence of K^+(Acosta-Cruz et al., 2012). The strains of *Pseudomonas solanacearum,* in association with plant roots, were able to survive for 24 days in the soil (Goddard et al., 2001). An introduced bacteria *Pseudomonas putida* GR12-2, isolated from the Arctic region and reported to secrete antifreeze protein in the growth medium, was growing and promoting root elongation of spring and winter canola at 5°C (Sun et al., 1995).

SUBSTRATE AVAILABILITY IN SOIL

Often the microorganisms face survival threat due to non-availability of substrates in the soil. Substrate availability of soil is defined by its biological carrying capacity or bioavailability in soil. The bioavailability of metals from soil depends on factors like CEC, organic matter content, content of clay minerals and hydrous metal oxides, pH, buffering capacity, redox potential, water content and temperature (Mugnier and Jung, 1985). The region outside rhizosphere majorly comprises of bulk soil and is considered nutrient-deficient (Kloepper, 1994). Bulk soil being oligotrophic in nature holds lesser microbial population with low species diversity. The reason behind bulk soil oligotrophy is the unavailability of carbon-rich nutrients in this region. The scarcity of nutrition may also be due to recalcitrance or location of nutrition at the sites not easily reachable by microorganisms. The availability of carbon is the deciding factor for the soil microbial population (Wardle, 1992). Root exudation by plants affects the carbon and nutrient availability of soil (Bowen and Rovira, 1991; Grayston et al., 1997). Carbon reaches soil from plants as litter material and root debris (Grayston et al., 1997). The plant growth stimulation through bacterial inoculants is affected by soil nutritional condition. The inoculants have abetter stimulatory effect on plant growth in nutrient-deficient soil than nutrient-rich one (Egamberdiyeva, 2007).

TESTING BACTERIAL SURVIVAL

A variety of methods are available for the quantification of survived microorganisms. Microbial activity can be measured in terms of biomass, respiration, and carbon flux through the biotic system. The standard microbial method is the most applied one in laboratories, which combines plate counting using selective or differential media for selecting antibiotic-resistant mutants. In many studies, rifampicin-resistant mutants were used (Dupler and Baker, 1984; Liu and Sinclair, 1992; Wiehe and Höflich, 1995; Nautiyal, 1997; Refshauge and Nayudu, 2001; Bolstridge et al., 2009). The possibility of antibiotic resistance trait loss of inoculant during the observation always exists, hence false results can be expected in such studies. Importantly, in most of the similar studies, control experiments on the fitness of the mutants derived from wild-type parents were not performed (Liang et al., 1982). In a study on *Bacillus cereus* UW85, spontaneous and permanent loss of streptomycin resistance was reported between 14-116 days of bacterial growth (Halverson et al., 1993). However, sometimes the antibiotic resistance loss could be temporary (Nairn and Chanway, 2002). Therefore, the use of double or triple antibiotic-resistant mutants was advocated to overcome the loss of single resistance and to increase the selectivity of plating method (Young and Burns, 1993). However, deliberate introduction of single or multiple antibiotic resistant traits may slow the growth rate of inoculant by interfering with their normal metabolism. In *Rhizobium* spp., decreased symbiotic effectiveness has been associated with induced streptomycin resistance (J.T. et al., 1992). Mutants of *Rhizobium* spp., resistant to both streptomycin and rifampin had significantly slower growth rates *in vitro* than wild-type strains (Liljeroth et al., 1991; J.T. et al., 1992).

For selective enumeration of inoculant population, enzyme-linked immunoassay (Rose et al., 2011) is a faster and less laborious method than plate counting (Schloter et al., 1995). Molecular methods, which are now being used more frequently in microbial studies, are highly specific and accurate in observations. Study on green fluorescent protein-tagged bacteria was initiated quite early for testing bioinoculant survival efficacy (Normander et al., 1999). By using molecular markers like green fluorescent protein or fluorescent antibodies, the location of individual rhizobacteria can be traced using laser confocal scanning microscopy (Bloemberg et al., 2000; Bloemberg and Lugtenberg, 2001; Bloemberg and Lugtenberg, 2001). The combined use of dilution plate counts, luminometry and charge-coupled device imaging shows the potential of luminescence-based detection as a powerful system for assessment of rhizosphere colonization by microbial inocula containing artificially inserted lux genes (Rattray et al., 1995). The use of rRNA-targeting probes for detecting the metabolic activity of microorganisms in the rhizosphere revealed that bacteria located at the root tip are metabolically more active compared to those surviving in other root regions (Lubeck et al., 2000).

Improving the Bioinoculant Survival in Soil

Strain Selection

Physiological characters of inoculant are the most significant set of deciding factors for bioinoculant survival. Since microbial physiology of microorganism is species-specific, selecting the appropriate strain is crucial and requires a better understanding of rhizosphere population dynamics. The inclusion of bacteria that are rapid colonizers and can outnumber the invading pathogenic microbial population is desirable in developing bioinoculants. Some plant species provide certain carbon and nitrogen-based nutrients, which can only be utilized by certain bacteria possessing the necessary plasmids for metabolism (Boivin et al., 1989). The phenomenon of antibiosis in the rhizosphere (Li and Alexander, 1990) or use of selective substrates like detergents (Devliegher et al., 1995) can be employed for retaining inoculant selectivity in the soil. Introduction of spore-formingPGPR can improve inoculant survival in soil. However, under nutritionally-stressed conditions, these bacteria turn dormant by producing endospores, demonstrating their non-usability during the period.

Certain PGPR (*Pseudomonas aeruginosa*, *Pseudomonas cepacia* and *Bacillus cereus*) may infect human beings as opportunistic pathogens. Therefore, the threats involved with their use should be clarified before large-scale acceptance, registration and adoption for pest and disease management (Nakkeeran et al., 2006). Prior understanding of rhizosphere ecology could be greatly advantageous for crop improvement. The knowledge of spatial-temporal dynamics of root pathogen will help in introducing the strains that may have similar dynamics for crop protection or phytostimulation. Rhizosphere bacteria can be characterized according to their colonization behavior. They may have specificity towards root colonization, and can therefore, be characterized either as root tip colonizer or root base colonizer (Liljeroth et al., 1991). However, the ecology of procured strains may not be available many times.

An application of plant-induced genes in rhizobacteria can help in their survival. This way, the bioinoculant activity could be expressed whenever required, thus avoiding the unnecessary metabolic load on the target cells. In a study on the effect of maize root exudates on transcriptomic profiling of *Bacillus amylolique faciens* FZB42, it was found that 86% of plant genes were up-regulated due to treatment. Several groups of the genes, which were strongly induced by the root exudates, were reported to be involved in metabolic pathways relating to nutrient utilization, bacterial chemotaxis and non-ribosomal synthesis of antimicrobial peptides and polyketides (Fan et al., 2012). A similar result was obtained from wheat seed extract on the genes of *Azospirillum brasilense* Sp245, which were identified with differential fluorescence induction (DFI) promoter trapping (Pothier et al., 2007). The use of promoters can aid in selective induction of beneficial genes. This approach is employable in biocontrol of microorganism-originated plant diseases because rhizosphere is often the site from where the plant infection starts.

Choice of Carrier Medium

Carrier formulation for increased bioinoculant survival is the primary focus of agricultural research. Unfortunately, efficient PGPRs may not always be good survivors in the environment. However, commercial use of PGPR requires the inoculum types that may retain high cell viability in the soil and can easily be transported and applied to the seeds. Hence, sometimes due to limitations in choice of proper inoculant it becomes necessary to utilize efficient carrier medium for supporting inocula in a hostile environment. In most bioinoculant studies, either liquid suspensions or bacteria incorporated into peat (Caesar and Burr, 1991) were used as carriers. Carriers help bioinoculants by providing a protective surface or optimal pore size for their protective micro-habitation and also by providing nutrient substrates. Artificial carriers immobilize the bacteria, fungi or enzymes and are made of organic polymers that form porous matrices around the microbial cell (Kierstan and Bucke, 1977). An ideal carrier favors inoculum survival, has long shelf life, is non-toxic to the environment and the inoculant, allows the timely release of inoculant and is compatible with other agrochemicals (Nakkeeran et al., 2005). Commercially available carrier materials are typically based on milled peat, clays, rice, bran, seeds, or other complex organic matrices (Yardin et al., 2000). Peat, vermiculite and talc are some of the carriers that have a shelf life of up to 12 months for non-sporing rhizobacteria like *Pseudomonas fluorescens* (Caesar and Burr, 1991). Calcium alginate, agarose and κ-carrageenan have been the material of choice for various reasons (J.T. et al., 1992). Such carriers are advantageous for inoculant survival because of their non-toxic nature, capacity to accommodate added compounds that can improve the physiology of cell, and relatively quick diffusion rate of gases, liquids or dissolved compounds. The application of these carriers in aqueous medium allows the relatively free exchange of gases by the diffusion of substrates and metabolic products through an aqueous phase.

Our greater understanding of soil ecology can enhance bioinoculant survival in the carrier medium. Heavily colonizing rhizobacterial types are the inoculant of choice, but if the microorganism is a slow grower, the multiple inoculations of the inoculants at targeted sites, with delayed delivery effect, is preferable. Inoculant delivery in the soil can be delayed by encapsulating the inoculant by spray-drying the cells with the carrier material of choice. Encapsulation increases the survival rate of bacteria, delays their delivery in the rhizosphere and gives inoculant the proper time as well as chance to colonize the root system at various loci in the soil (J.T. et al., 1992). The calcium alginate beads and skimmed milk are the best carrier choices for encapsulation. Survival of inoculant has been prolonged by using dry alginate-beads (Bashan and Gonzalez, 1999). Alginate beads delay the release of inoculant in the soil and hence, can control the sharp population decline of inoculant (Bashan, 1986). In the study on *Pseudomonas fluorescens* a relationship was found between bacterial survival and residual moisture in the microspheres. The optimal survival of the stored bacteria was observed at 25% residual moisture (Amiet-Charpentier et al., 1998). Looking at the commercial aspect of the carrier, use of peat is also preferable. It has been successfully used to enhance the survival of *Rhizobium* (Höflich et al., 1995), *Pseudomonas* (Höflich et al., 1995; Rabindran and Vidhyasekaran, 1996) and *Bacillus subtilis*(Kinsella et al., 2009)-based inoculants. Different organic and inorganic carriers used in formulating PGPR isolates using solid or liquid fermentation technologies have been delivered through seed treatment, seed bio-priming, root dipping of seedlings, soil application, foliar spray, fruit spray, hive insert or sucker treatment (Nakkeeran et al., 2006).

Conclusion

The effect of soil physiology and ecology on introduced microorganisms is not properly understood yet, and requires further study. The key to successful colonization of bioinoculants lies in the proper understanding of soil type, rhizosphere and selection of appropriate strains. Results suggest that field trials conducted without considering soil quality factors and carrier delivery systems were of dubious value when evaluated for bioinoculant performance (Horwath et al., 1998). The genetically modified rhizobacteria require even greater consideration because they may not have the same adaptability as their wild parents (van Elsas et al., 1991).

Climatic conditions of the geographical regions are varied, difficult to change and often hostile to the introduced bioinoculants. The use of adaptive bioinoculant strains is desirable for harsh climates. Inserting specific genes in the bioinoculant or inducing some extra metabolic pathways necessary to combat the extreme environment may help them cope with unfavorable environments. Moreover, the plant growth-promoting traits of the bioinoculant can be transferred to native soil bacteria by making them competent. Slow-growing or weaker inoculants can be established in the rhizosphere, either by using protective carrier material or by temporarily altering the buffering mechanism of root micro-ecosystem.

An estimate shows that as much as 99% or more of microbial life remains 'unculturable' and, therefore, cannot be studied and understood in a way that microbial ecologists have become accustomed to, over the past century (Leveau, 2007). In the light of this fact, the role of metagenomics in rhizosphere study cannot be neglected. In one interesting study, it was found that the use of rifampicin-resistant *P. fluorescens* R2f Rp^r on loamy sand led to increase in their non-culturable counterpart (van Overbeek et al., 1995). Further study of the rhizosphere by comparative metagenomics could reveal several important answers related to the role of the unculturable fraction of the rhizosphere community (Leveau, 2007).

Introduction of a microbial mixture in the soil for microbial colonization is a subject of attention in soil microbiology because one species system is often vulnerable to competitors, pathogens or other factors (Whipps, 2001). More studies are required on co-inoculant PGPRs or a mixture of strains for ensuring their successful root colonization (Goddard et al., 2001). The co-inoculants should contain two or more ecologically distinct bacteria for coping with the negative effects of ecological diversity in the rhizosphere. In this way, beneficial bacteria can be expressed more continually, independent of ecological differences. The combination of diazotrophs with hydrolytics or producers of growth stimulators and phytohormones with phosphate mobilizing bacteria can be used as an example of artificial microbial associations.

REFERENCES

Abat, B. (2006). *Growth of agriculturally important Pseudomonas spp. and Azotobacterchroococcum on beer waste and observation of their survival in peat:* Middle East Technical University. 91 p.

Acea, M. J., Moore, C. R., Alexander, M. (1988). Survival and growth of bacteria introduced into soil. *Soil Biology and Biochemistry* 20: 509-515.

Acosta-Cruz, E., Wisniewski-Dye, F., Rouy, Z., Barbe, V., Valdes, M., Mavingui, P. (2012). Insights into the 1.59-Mbp largest plasmid of *Azospirillum brasilense* CBG497. *Archives of microbiology* 194: 725-736.

Afzal, M., Yousaf, S., Reichenauer, T. G., Kuffner, M., Sessitsch, A. (2011). Soil type affects plant colonization, activity and catabolic gene expression of inoculated bacterial strains during phytoremediation of diesel. *Journal of Hazardous Materials* 186: 1568-1575.

Amiet-Charpentier, C., Gadille, P., Digat, B., Benoit, J. P. (1998). Microencapsulation of rhizobacteria by spray-drying: formulation and survival studies. *Journal of microencapsulation* 15: 639-659.

Anburaj, R., Nabeel, M. A., Sivakumar, T., Kathiresan, K. (2012). The role of rhizobacteria in salinity effects on biochemical constituents of the halophyte Sesuvium portulacastrum. *Russ J Plant Physiol* 59: 115-119.

Andrade, G., Mihara, K. L., Linderman, R. G., Bethlenfalvay, G. J. (1998). Soil aggregation status and rhizobacteria in the mycorrhizosphere. *Plant and Soil* 202: 89-96.

Bais, H. P., Weir, T. L., Perry, L. G., Gilroy, S., Vivanco, J. M. (2006). The role of root exudates in rhizosphere interactions with plants and other organisms. *Annual Review of Plant Biology* 57: 233-266.

Baldani, J. I., Weaver, R. W. (1992). Survival of clover rhizobia and their plasmid-cured derivatives in soil under heat and drought stress. *Soil Biology and Biochemistry* 24: 737-742.

Bashan, Y. (1986). Alginate beads as synthetic inoculant carriers for slow release of bacteria that affect plant growth. *Applied and Environmental Microbiology* 51: 1089-1098.

Bashan, Y. (1986). Enhancement of wheat root colonization and plant development by *Azospirillum brasilense* Cd. following temporary depression of rhizosphere microflora. *Applied and Environmental Microbiology* 51: 1067-1071.

Leveau, J. H. J. (2007). The magic and menace of metagenomics: prospects for the study of plant growth-promoting rhizobacteria. *European Journal of Plant Pathology* 119: 279-300.

Li, D.-M., Alexander, M. (1990). Factors affecting co-inoculation with antibiotic-producing bacteria to enhance rhizobial colonization and nodulation. *Plant and Soil* 129: 195-201.

Liang, L. N., Sinclair, J. L., Mallory, L. M., Alexander, M. (1982). Fate in Model Ecosystems of Microbial Species of Potential Use in Genetic Engineering. *Applied and Environmental Microbiology* 44: 708-714.

Liljeroth, E., Burgers, S. L. G. E., Veen, J. A. (1991). Changes in bacterial populations along roots of wheat (*Triticum aestivum* L.) seedlings. *Biology and Fertility of Soils* 10: 276-280.

Liu, X., Bimerew, M., Ma, Y., Muller, H., Ovadis, M., Eberl, L., Berg, G., Chernin, L. (2007). Quorum-sensing signaling is required for production of the antibiotic pyrrolnitrin in a rhizospheric biocontrol strain of *Serratia plymuthica*. *FEMS microbiology letters* 270: 299-305.

Liu, Z. L., Sinclair, J. B. (1992). Population dynamics of *Bacillus megaterium* strain B153-2-2 in the rhizosphere of soybean. *Phytopathology* 82: 1297-1301.

Loper, J. E., Haack, C., Schroth, M. N. (1985). Population dynamics of soil pseudomonads in the rhizosphere of potato (*Solanum tuberosum* L.). *Applied and Environmental Microbiology* 49: 416-422.

Lubeck, P. S., Hansen, M., Sorensen, J. (2000). Simultaneous detection of the establishment of seed-inoculated *Pseudomonas fluorescens* strain DR54 and native soil bacteria on sugar beet root surfaces using fluorescence antibody and in situ hybridization techniques. *FEMS Microbiology Ecology* 33: 11-19.

Lucas, R. E., Davis, J. F. (1961). Relationships between pH values of organic soils and availabilities of 12 plant nutrients. *Soil Science* 92: 177-182.

Lugtenberg, B. J., Dekkers, L., Bloemberg, G. V. (2001). Molecular determinants of rhizosphere colonization by *Pseudomonas*. *Annual review of phytopathology* 39: 461-490.

Lynch, J. M. (1990). The potential for gene exchange between rhizosphere bacteria. In: Fry J, Day M, (Eds.). *Bacterial Genetics in Natural Environments*: Springer Netherlands. pp. 172-181.

Mateille, T., Fould, S., Dabiré, K. R., Diop, M. T., Ndiaye, S. (2009). Spatial distribution of the nematode biocontrol agent Pasteuria penetrans as influenced by its soil habitat. *Soil Biology and Biochemistry* 41: 303-308.

Mathre, D. E., Cook, R. J., Callan, N. W. (1999). From discovery to use: Traversing the world of commercializing biocontrol agents for plant disease control. *Plant Disease* 83: 972-983.

Mongodin, E. F., Shapir, N., Daugherty, S. C., DeBoy, R. T., Emerson, J. B., Shvartzbeyn, A., Radune, D., Vamathevan, J., Riggs, F., Grinberg, V., Khouri, H., Wackett, L. P., Nelson, K. E., Sadowsky, M. J. (2006). Secrets of soil survival revealed by the genome sequence of *Arthrobacter aurescens* TC1. *PLoS Genet* 2: e214.

Mugnier, J., Jung, G. (1985). Survival of bacteria and fungi in relation to water activity and the solvent properties of water in biopolymer gels. *Applied and Environmental Microbiology* 50: 108-114.

Nairn, J. D., Chanway, C. P. (2002). Temporary loss of antibiotic resistance by marked bacteria in the rhizosphere of spruce seedlings. *FEMS Microbiology Reviews* 40: 167-170.

Nakkeeran, S., Fernando, W. G. D., Siddiqui, Z. A. (2005). Plant growth promoting rhizobacteria formulationsand its scopeincommercializationfor the management ofpests anddiseases. In: Siddiqui ZA, (Ed.). *PGPR: Biocontrol and Biofertilization*. The Netherlands Springer.

Nakkeeran, S., Fernando, W., Siddiqui, Z. (2006). Plant growth promoting rhizobacteria formulations and its scope in commercialization for the management of pests and diseases *PGPR: Biocontrol and biofertilization*. In: Siddiqui Z, (Ed.): Springer Netherlands. pp. 257-296.

Nautiyal, C. S. (1997). Rhizosphere competence of *Pseudomonas* sp. NBRI9926 and *Rhizobium* sp. NBRI9513 involved in the suppression of chickpea (*Cicer arietinum* L.) pathogenic fungi. *FEMS Microbiology Ecology* 23: 145-158.

Normander, B., Hendriksen, N. B., Nybroe, O. (1999). Green fluorescent protein-marked *Pseudomonas fluorescens*: localization, viability, and activity in the natural barley rhizosphere. *Applied and Environmental Microbiology* 65: 4646-4651.

Or, D., Smets, B. F., Wraith, J. M., Dechesne, A., Friedman, S. P. (2007). Physical constraints affecting bacterial habitats and activity in unsaturated porous media – a review. *Advances in Water Resources* 30: 1505-1527.

Perotti, E. B., Menendez, L. T., Gaia, O. E., Pidello, A. (2005). *Pseudomonas fluorescens* survival in soils with different contents of organic matter. *Revista Argentina de microbiologia* 37: 102-105.

Perotti, E., Chapo, G., Menendez, L., Pidello, A. (1985). Survival of *Azospirillum* in the rhizosphere of *Festuca arundinacea*. *Revista Argentina de microbiologia* 17: 103-108.

Pothier, J. F., Wisniewski-Dyé, F., Weiss-Gayet, M., Moënne-Loccoz, Y., Prigent-Combaret, C. (2007). Promoter-trap identification of wheat seed extract-induced genes in the plant-growth-promoting rhizobacterium *Azospirillum brasilense* Sp245. *Microbiology* 153: 3608-3622.

Raaijmakers, J. M., Sluis, L. v. d., Bakker, P. A. H. M., Schippers, B., Koster, M., Weisbeek, P. J. (1995). Utilization of heterologous siderophores and rhizosphere competence of fluorescent *Pseudomonas* spp. *Canadian journal of microbiology* 41: 126-135.

Rabindran, R., Vidhyasekaran, P. (1996). Development of a formulation of *Pseudomonas fluorescens* PfALR2 for management of rice sheath blight. *Crop Protection* 15: 715-721.

Rainey, P. B. (1999). Adaptation of *Pseudomonas fluorescens* to the plant rhizosphere. *Environmental Microbiology* 1: 243-257.

Rattray, E., Prosser, J., Glover, L., Killham, K. (1995). Characterization of rhizosphere colonization by luminescent *Enterobacter cloacae* at the population and single-cell levels. *Applied and environmental microbiology* 61: 2950-2957.

Refshauge, S. J., Nayudu, M. (2001). Invasion by *Pseudomonas syringae* pv. coriandricola is responsible for bacterial blight of coriander. *Plant Science* 161: 621-625.

Rhoades, J. D. (1982). Cation Exchange Capacity. *Methods of Soil Analysis Part 2 Chemical and Microbiological Properties*: 149-157.

Roberts, D. P., Yucel, I., Larkin, R. P. (1998). Genetic approaches for analysis and manipulation of rhizosphere colonization by bacterial biocontrol agents. In: Boland GJ, Kuykendall LD, (Eds.). *Plant Microbe Interactions and Biological Control*. New York, USA: Marcel Dekker Inc. pp. 415-431.

Rose, M., Deaker, R., Potard, S., Tran, C., Vu, N., Kennedy, I. (2011). The survival of plant growth promoting microorganisms in peat inoculant as measured by selective plate counting and enzyme-linked immunoassay. *World Journal of Microbiology and Biotechnology* 27: 1649-1659.

Schloter, M., Assmus, B., Hartmann, A. (1995). The use of immunological methods to detect and identify bacteria in the environment. *Biotechnology Advances* 13: 75-90.

Shakir, M. A., Bano, A., Arshad, M. (2011). Short Communication: Rhizosphere bacteria containing ACC-deaminase conferred drought tolerance in wheat grown under semi-arid climate. *Soil & Environment* 31.

Soby, S., Bergman, K. (1983). Motility and chemotaxis of *Rhizobium meliloti* in soil. *Applied and Environmental Microbiology* 46: 995-998.

Strigul, N. S., Kravchenko, L. V. (2006). Mathematical modeling of PGPR inoculation into the rhizosphere. *Environmental Modelling & Software* 21: 1158-1171.

Sun, X., Griffith, M., Pasternak, J. J., Glick, B. R. (1995). Low temperature growth, freezing survival, and production of antifreeze protein by the plant growth promoting rhizobacterium *Pseudomonas putida* GR12-2. *Canadian journal of microbiology* 41: 776-784.

Sylvia, D. M., Fuhrmann, J. J., Hartel, P. G., Zuberer, D. A. (2004). *Principles and Applications of Soil Microbiology*: Prentice Hall. 672 p.

Thompson, I. P., Young, C. S., Cook, K. A., Lethbridge, G., Burns, R. G. (1992). Survival of two ecologically distinct bacteria (*Flavobacterium and arthrobacter*) in unplanted and rhizosphere soil: Field studies. *Soil Biology and Biochemistry* 24: 1-14.

Turnbull, G. A., Morgan, J. A. W., Whipps, J. M., Saunders, J. R. (2001). The role of motility in the *in vitro* attachment of *Pseudomonas putida* PaW8 to wheat roots. *FEMS Microbiology Ecology* 35: 57-65.

Tyler, H. L., Triplett, E. W. (2008). Plants as a habitat for beneficial and/or human pathogenic bacteria. *Annual review of phytopathology* 46: 53-73.

van Elsas, J. D., van Overbeek, L. S., Feldmann, A. M., Dullemans, A. M., de Leeuw, O. (1991). Survival of genetically engineered *Pseudomonas fluorescens* in soil in competition with the parent strain. *FEMS microbiology letters* 85: 53-64.

van Overbeek, L. S., Eberl, L., Givskov, M., Molin, S., van Elsas, J. D. (1995). Survival of, and induced stress resistance in, carbon-starved *Pseudomonas fluorescens* cells residing in soil. *Applied and Environmental Microbiology* 61: 4202-4208.

Vandenhove, H., Merckx, R., Wilmots, H., Vlassak, K. (1991). Survival of *Pseudomonas fluorescens* inocula of different physiological stages in soil. *Soil Biology and Biochemistry* 23: 1133-1142.

Wardle, D. A. (1992). A comparative assessment of factors which influence microbial biomass carbon and nitrogen levels in soil. *Biological Reviews* 67: 321-358.

Wessendorf, J., Lingens, F. (1989). Effect of culture and soil conditions on survival of *Pseudomonas fluorescens* R1 in soil. *Applied Microbiology and Biotechnology* 31: 97-102.

Whipps, J. M. (2001). Microbial interactions and biocontrol in the rhizosphere. *Journal of experimental botany* 52: 487-511.

Wiehe, W., Höflich, G. (1995). Establishment of plant growth promoting bacteria in the rhizosphere of subsequent plants after harvest of the inoculated precrops. *Microbiological Research* 150: 331-336.

Williams, A. P., Avery, L. M., Killham, K., Jones, D. L. (2007). Survival of *Escherichia coli* O157:H7 in the rhizosphere of maize grown in waste-amended soil. *Journal of applied microbiology* 102: 319-326.

Wright, D. A., Killham, K., Glover, L. A., Prosser, J. I. (1995). Role of pore size location in determining bacterial activity during predation by protozoa in soil. *Applied and Environmental Microbiology* 61: 3537-3543.

Yardin, M. R., Kennedy, I. R., Thies, J. E. (2000). Development of high quality carrier materials for field delivery of key microorganisms used as bio-fertilisers and bio-pesticides. *Radiation Physics and Chemistry* 57: 565-568.

Young, C. S., Burns, R. G. (1993). Detection, survival, and activity of bacteria added to soil. *Soil biochemistry* 8: 1-63.

Young, C. S., Lethbridge, G., Shaw, L. J., Burns, R. G. (1995). Survival of inoculated *Bacillus cereus* spores and vegetative cells in non-planted and rhizosphere soil. *Soil Biology and Biochemistry* 27: 1017-1026.

Zachow, C., Grosch, R., Berg, G. (2011). Impact of biotic and a-biotic parameters on structure and function of microbial communities living on sclerotia of the soil-borne pathogenic fungus *Rhizoctonia solani*. *Applied Soil Ecology* 48: 193-200.

In: Recent Advances in Microbiology. Volume 3 ISBN: 978-1-53614-057-6
Editors: N.C. Gautam, S.P. Tiwari and R. Sharma

Chapter 5

ANTIMICROBIAL INHERITANCE OF MILK AND MILK PRODUCTS

Dinesh Chandra Rai[1,*] and Tanweer Alam[2]
[1]Department of A. H. & Dairying, Institute of Agricultural Sciences,
Banaras Hindu University, Varanasi, UP, India
[2]Indian Institute of Packaging, Chennai, India

INTRODUCTION

Milk is defined as the white lacteal secretion obtained by complete milking from the herd of healthy milch animals, five days after calving and 15 days before calving. Milk is valued because it is an important source of many of the nutrients essential for the proper development and maintenance of the human body, and one of the most nutritious foods. It is rich in high quality protein providing all essential amino acids. It contributes to total daily energy intake as well as essential fatty acids, immunoglobulins, and other micronutrients. It is the primary source of nutrition for young mammals before they are able to digest other types of food. First lactation is known as colostrum, which carries the mother's antibodies to the baby, improving the immunity of neonates and preventing the risk of many diseases. Cow's milk is the predominant type of milk consumed in several countries, although goat, buffalo, sheep, and camel milk are also consumed.

Wide ranges of milk products are available throughout the world, including curd, cheese, milk powder, ice cream, butter, fermented products etc. The precise nutrient composition of raw milk varies from species to species, and by a number of other factors, but it contains significant amounts of saturated fat, protein and calcium as well as vitamin

* Corresponding Author E-Mail: dcrai.bhu@gmail.com.

C. A recent study states that approximately 73% of the calcium, 33% of the phosphorus, 31% of the riboflavin, 21% of the vitamin B12, 19% of the protein, 16% of the magnesium, 17% of the vitamin A, 10% of the vitamin B6, and 6% of the thiamin in addition to appreciable amounts of vitamin D and niacin is contributed by dairy-related food,. Thus, milk and dairy products act as nutrient-dense foods providing abundant amounts of protein, vitamins and minerals necessary for growth and development. A number of proteins found in milk under various conditions exhibit antimicrobial activity. For example, immunoglobulins (antibodies) are protective proteins that are important in the transfer of passive immunity from the mother to the neonate. The immunoglobulins protect the neonate from infection until their own immune system is developed. Immunoglobulins are a component of the natural defense mechanism. They are synthesized in response to the presence of foreign particles, such as bacteria and viruses. The synthesis process is specific for the foreign particle that is present, resulting in an immunoglobulin structure that is capable of recognizing the foreign particle and eliminating it from the body. Immunoglobulins are found in high concentrations in colostrum, the first milk, and in low concentrations in milk. In addition to its value as a nutrient source, the purpose of this chapter is to critically review the antimicrobial properties of milk and milk products.

Keywords: milk, milk composition, nutritional composition, milk products

Table 1. Composition of Milk from Different Animal Species (Amount per 100 g)

Species	Percentage of Composition				
	Water	Fat	Protein	Lactose	Ash
Ass	90.0	1.3	1.7	6.5	0.5
Buffalo	84.2	6.6	3.9	5.2	0.8
Camel	86.5	3.1	4.0	5.6	0.8
Cat	84.6	3.8	9.1	4.9	0.6
Cow	86.6	4.6	3.4	4.9	0.7
Dog	75.4	9.6	11.2	3.1	0.7
Elephant	67.8	19.6	3.1	8.8	0.7
Ewe	79.4	8.6	6.7	4.3	1.0
Goat	86.5	4.5	3.5	4.7	0.8
Guinea pig	82.2	5.5	8.5	2.9	0.9
Human	87.7	3.6	1.8	6.8	0.1
Llama	86.5	3.2	3.9	5.6	0.8
Mare	89.1	1.6	2.7	6.1	0.5
Porpoise	41.1	45.8	11.2	1.3	0.6
Reindeer	68.2	17.1	10.4	2.3	1.5
Sow	89.6	4.8	1.3	3.4	0.9
Whale	70.1	19.6	9.5	--	1.0

Source: Dairy India Year Book – 2007.

Table 2. Composition of Milk from Different Breeds of Animal (Amount per 100 g)

Species	Percentage of Composition				
	Water	**Fat**	**Protein**	**Lactose**	**Ash**
Cow Holstein	87.74	3.40	3.22	4.87	0.68
Shorthorn	87.19	3.94	3.32	4.99	0.70
Ayreshire	87.10	4.00	3.58	4.67	0.68
Brown Swiss	86.59	4.01	3.61	5.04	0.73
Guernsey	85.39	4.95	3.91	4.93	0.74
Jersey	85.09	5.37	3.92	4.93	0.71
Sindhi	86.07	4.90	3.42	4.91	0.70
Gir	86.44	4.73	3.32	4.85	0.66
Tharparkar	86.58	4.55	3.36	4.83	0.68
Sahiwal	86.42	4.55	3.33	5.04	0.66
Buffalo Murrah	83.63	6.56	3.88	5.23	0.70

Source: Dairy India Year Book – 2007.

MILK

Milk is a complex, nutritious product that contains more than 100 substances that are either in solution, suspension or emulsion in water. For example:

- Casein, the major protein of milk, is dispersed as a great number of solid particles so tiny that they do not settle, but remain in suspension. These particles are called micelles, and the dispersion of the micelles in the milk is referred to as a colloidal suspension;
- The fat and fat soluble vitamins in the milk are in the form of an emulsion; that is, a suspension of small liquid globules that do not mix with the water in milk;
- Lactose (milk sugar), some proteins (whey protein), mineral salts and other substances are soluble; that is, they are entirely dissolved in the water in milk.

Table 3. Mineral and Vitamin Concentrations in Milk (mg/100 ml)

Minerals mg/100 ml	**Vitamins µg/100 ml1**
Potassium 138	Vit. A 30.0
Calcium 125	Vit. D 0.06
Chloride 103	Vit. E 88.0
Phosphorus 96	Vit. K 17.0
Sodium 58	Vit. B1 37.0
Sulfur 30	Vit. B2 180.0
Magnesium 12	Vit. B6 46.0
Trace minerals 2 < 0.1	Vit. B12 0.42
	Vit. C 1.7

Source: Dairy Directory Year Book: (2006).

The casein micelles and the fat globules give milk most of its physical characteristics, and give taste and flavor to dairy products such as butter, cheese, yogurt, etc. The composition of milk varies considerably with the breed of animals, stage of lactation, feed, season of the year, and many other factors. However, some relationships between constituents are very stable and can be used to indicate whether any tampering with the milk composition has occurred. For example, milk of normal composition has a specific gravity that varies normally from 1.023 to 1.040 (at 20°C) and a freezing point that varies from -0.518 to -0.534°C. Any alteration, by addition of water for example, can be easily identified because these characteristics of milk will no longer be in the normal range. Extremes of temperature, acidity (0.14-0.16), pH (6.6-6.7) or contamination by microorganisms can rapidly decrease its quality. Milk is a highly perishable product that should be chilled to 4±1°C as soon as possible after collection to check the augmentation of microorganism and thus, to enhance the longevity of milk.

Nutrient Composition of Milk

Lactose

Lactose is the principal and typical carbohydrate of milk, known as milk sugar. Glucose, galactose and other sugars such as oligosaccharides are also present in traces. Lactose exists in true solution in milk. It is a disaccharide, composed of two molecules of monosaccharides, glucose and galactose. Lactose is readily fermented by the lactic acid fermenting bacteria producing lactic acid and has significance in milk and milk products. It exists in two isomeric forms, designated as α and ß forms, of which the ß form is more soluble than the alpha form in water. Lactose content of cow milk is 4.9 percent. The lactose content of milk is inversely proportional to the ash content of the milk. Udder infection promotes an increased level of chloride in the milk and depresses the secretion of lactose. The metabolism of lactose in the intestine is considerably slower and hence, it passes through the intestine slowly where it promotes the growth of lactic acid fermenting bacteria. The acidic condition produced due to fermentation of lactose to lactic acid increases the solubility of lactose and improves its absorption by forming a soluble complex with calcium. Slow absorption of lactose from the intestine, coupled with low pH, induces laxative effect by increasing peristalsis. Even on a higher lactose diet, the lactose concentration does not shoot up. Hence, diabetic persons can safely consume milk or milk products rich in lactose without any adverse impact.

Proteins

Casein

Casein is one of the major milk proteins, which is precipitated at pH 4.6 and is exclusively present in milk. It is present in spherical bodies called micelles, which vary in size with negative surface charge. The caseins of milk may be sub-divided into five main classes, αs1, αs2, ß, gamma and k-caseins. In milk, casein is present in combination with calcium in the form of calcium hydrogen caseinate.

Whey Proteins

Whey proteins are obtained during milk precipitation (pH 4.6) of casein. These are the alpha-lactalbumin, beta-lactoglobulin, immunoglobulins, lactoferrin, transferrin, and proteose-peptone fractions. Most of these are globular proteins subject to heat denaturation. a-lactalbumin exists in two variant forms A and B, and is susceptible to denaturation by unfolding of the tertiary structure. ß-lactoglobulin is identical to blood globulin, insoluble in water and is responsible for the transfer of antibodies. Normal milk contains 0.1%, whereas colostrum contains 6% of ß-lactoglobulin. They also contain 51% essential amino acids when compared to 45% in casein. Sulfur-containing amino acids, which are considered essential and important, are found in higher concentration in whey protein than in casein. Usually, the qualities of egg proteins are regarded as very high. However, the net protein utilization, biological value and the protein efficiency ratio of milk protein closely compete with the quality of egg protein. Lactalbumin is a whey protein, whose biological value, net protein utilization and protein efficiency ratio is considered superior when compared to the major milk protein, casein. Normally, double the quantum of vegetable protein is required to meet the daily requirement of essential amino acids when compared to that obtained from the milk proteins. Milk-derived peptides have functional properties like antimicrobial protein.

Table 4. Protein Concentration and Casein fractions in Milk

Sr. No	Protein and protein fractions	Concentration in milk g/L
1.	Total protein	36
2.	Casein	29.5
3.	alpha s1-Casein	11.9
4.	alpha s2-Casein	3.1
5.	beta Casein	9.8
6.	kappa Casein	3.5
7.	gamma Casein	1.2

Source: From Jensen, R. G. Ed., Handbook of Milk Composition, Academic Press, New York, p. 465, 1995.

lysozyme may have an indirect effect on the defense systems of hosts as an immunomodulator, through the stimulation of the immune system by breaking down products from the hydrolysis of peptidoglycan. This hypothesis has been supported by the findings that feeding infants lysozyme-enriched formulas results in an increased level of secretory IgA in feces. Lysozyme may also be responsible for some false positive results for antibiotics in milk. However, such milk is likely to be of high somatic cell count.

Lactoferrin

Lactoferrin (Lf), an iron chelating protein, occurs in many exocrine secretions of mammals. Lactoferrin, an iron-binding glycoprotein, was first isolated from cow's milk and subsequently from human milk. Lactoferrin is present in large quantities in mammalian secretions such as milk, tears, saliva, and seminal fluid, as well as in some white blood cells. It is one of the minor proteins naturally occurring in cow milk, at an average concentration of about 0.2 grams/liter. It is closely related to transferrin, a soluble glycoprotein that binds two iron ions (III) per molecule. Lf is also related to an iron chelating protein found in hen egg white, ovotransferrin or conalbumin. All three iron-containing proteins have similar structures and biological properties. They consist of a single polypeptide chain of molecular weight 80,000 daltons, to which one or two carbohydrate chains are attached. The molecules have two symmetrical parts, and each has one site for the binding of a ferric ion. For each ferric ion that is bound, one bicarbonate ion is also bound. Lf has been found in milk of some 12 species. In humans, it also has been shown to be present in most exocrine secretions, including bronchial mucus, tears, saliva, nasal secretions, cervical mucus and gastric juice. In human milk, Lf is one of the major whey proteins, constituting 10-30% of the total protein content. Nagasawa et al. (1972) have reported levels of Lf in human colostrum and milk of 4.9 and 1.6 mg/ml, respectively. In bovine milk, the levels are substantially lower, i.e., about 1 mg/ml and 0.2 mg/ml in colostrums and milk, respectively. However, the concentration of lactoferrin in bovine milk increases significantly with inflammatory conditions of the mammary glands, e.g., mastitis. In the native state, Lf in only partially saturated with iron. In human milk, which contains 0.3 to 0.5 μg/ml Fe, only 2-4% of this is bound to Lf. A major part of the Fe in milk is found in the fat phase, mainly as a component of xanthine oxidase. The rest is present in the form of low molecular complexes. A number of investigations have been carried out to elucidate the biological role of Lf. Although a number of activities have been demonstrated *in vitro*, the *in vivo* function of Lf is not well understood. However, two main roles for LF have been postulated - its antibacterial effect and its role in iron metabolism. The antibacterial effect of Lf under *in vitro* conditions is well established. The antibacterial mechanism has been assigned to the ability to chelate iron and deprive bacteria of the iron that is essential for growth.

In colostrum, the lactoferrin content can be as high as 0.5 to 1 grams/liter. During the dry period, lactoferrin concentration in mammary secretions from dry cows increases until about 30 days after drying off. The highest lactoferrin concentration found in cow mammary secretions is about 50 to 100 grams/liter. In human milk and colostrum, the reported concentrations of lactoferrin are 2 to 4 grams/liter and 6 to 8 grams/liter, respectively. In its natural state, lactoferrin is only partly saturated with iron (5 to 30%). Lactoferrin has many proposed biological functions, including antibacterial/ anti-inflammatory activities, defense against gastro-intestinal infections, participation in local secretory immune systems in synergism with some immunoglobulins and other protective proteins, provision of an iron-binding antioxidant protein in tissues, and possibly promotion of growth of animal cells such as lymphocytes and intestinal cells. The role for milk lactoferrin in iron absorption by the intestine has long been postulated, but remains unproven. Most micro-organisms need iron for growth and lactoferrin has the potential to inhibit the growth of bacteria, and even kill them by depriving them of iron. The effectiveness of the antibacterial activity of lactoferrin depends on the iron requirement of the organism, the availability of exogenous iron, and the concentration and degree of iron-saturation of lactoferrin. It has been shown that 'natural' lactoferrin is bacteriostatic against a wide range of micro-organisms, including Gram-negative bacteria with high iron requirements (coliforms, which are major mastitis pathogens), and also against some Gram-positive organisms such as *Staphylcoccus aureus* (also a major mastitis pathogen), bacillus species, and *Listeria monocytogenes*. Lactic acid bacteria in the stomach and intestine have low iron requirements and are generally not affected. There is also evidence that on certain streptococcal mutants and *Vibrio cholerae*, lactoferrin can exert a direct, bactericidal effect that is independent of iron-deprivation.

Lactoperoxidase

Peroxidase enzymes can kill bacteria by oxidative mechanisms. Peroxidase activity occurs in various exocrine gland secretions including saliva, tears, bronchial, nasal, and intestinal secretions, as well as in milk. Milk peroxidase is known as lactoperoxidase, which is one of the non-immunoglobulin protective proteins and a prominent enzyme that plays a role in protection against microbial invasion of the mammary gland. Each lactoperoxidase molecule contains one iron atom. Bovine milk contains concentrations of about 0.03 grams/liter. In bovine colostrum, the lactoperoxidase content is very low, but increases rapidly 4 to 5 days postpartum. The level of lactoperoxidase activity in human milk is about 20 fold lower than that in bovine milk. Lactoperoxidase itself has no antibacterial activity. However, together with hydrogen peroxide and thiocyanate, lactoperoxidase forms a potent natural antibacterial system, the so-called lactoperoxidase system. Both hydrogen peroxide and thiocyanate are naturally distributed in animal and human tissues, although they generally occur in very low concentrations. The antibacterial effect of the lactoperoxidase system is mediated by the reaction of hydrogen peroxide and thiocyanate under lactoperoxidase catalysis and the resultant generation of short-lived hypothiocyanate, which is thought to be a major antibacterial substance. The antibacterial property of the lactoperoxidase system is based upon inhibition of vital bacterial metabolic enzymes brought on by their oxidation by hypothiocyanate. Lactoperoxidase the protein has been discussed previously in the introductory section on inhibitors in milk. Lactoperoxidase is a basic (it is negatively charged below its isoelectric point) glycoprotein that has a molecular weight of 78,000 and contains one heme group. The enzyme has iron content between 0.068-0.071% and a carbohydrate content ranging from 9.9 - 10.2%. The identity and type of binding of the heme moiety has been investigated. Initially, it was proposed that the heme group was covalently bound to the protein moiety *via* an ester or amide bond. Scientists have, however, shown that no covalent heme-protein bond exists and that the prosthetic group is protoheme IX. Peroxidases catalyze reactions in which hydrogen peroxide is reduced and an appropriate electron donor is subsequently oxidized. A wide variety of organic and inorganic substances can serve as electron donors, but substrate specificity varies between various peroxidases. LP itself has no antibacterial effect, but in combination with certain co-factors, thiocyanate (SCN-) and hydrogen peroxide (H_2O_2) forms a potent antimicrobial system. The mechanism of this inhibition is largely understood as a result of studies on the inhibition of acid production by lactococci in cheese manufacture. The antibacterial mechanism is caused by the oxidation of vital SH-groups by OSCN-/ O2SCN- in critically significant metabolic enzymes, e.g., hexokinase and/or depletion of reduced nicotinamide adenine nucleotides. A wide variety of bacteria are influenced by the LP-system. With some bacteria, the effect is reversible, i.e., the bacteria are inhibited for a certain period of time; for others, the effect is bactericidal. Many Gram positive bacteria such as lactococci and lactobacilli are inhibited while many Gram-negative bacteria such as *Escherichia coli*, *Pseudomonas* spp. and *Salmonella* spp. are killed. Mammalian cells do not appear to be adversely affected by the LP-system. The LP-system did not have any mutagenic effects when tested against several mutagen-sensitive strains of *Salmonella typhimurium*. The *in vivo* activity of the LP system is not fully known. In bovine milk, LP is always present in excess. Buffalo's milk, ewe's milk, and goat's milk also contain high levels of LP. The factors limiting the activity of the LP system in milk from these species are the low concentrations of thiocyanate and hydrogen peroxide present. Some thiocyanate is normally present in milk, although the level is variable, i.e., between 1 to 15 mg/l. There is some evidence of *in vivo* production of hydrogen peroxide by lactic acid bacteria. Most investigations of the antibacterial activity of the LP system have been carried out *in vitro*. In these experiments, LP, thiocyanate and hydrogen peroxide are added to a suitable medium. In milk, the system can be "activated" by the addition of thiocyanate and hydrogen peroxide or an appropriate hydrogen peroxide source. There are several methods for determining the concentration and antimicrobial activity of the LP system.

Scientific study evidenced that late weaning may be a desirable way to protect fattening rabbits against colibacillosis. Indeed, it has been reported by imminent scientists that suckling rabbits are protected against infection with an EPEC strain that belongs to the O103 sero-group that is known to be highly virulent (Nattress & Baker, 2003). When weaned rabbits are infected by using similar conditions (infection with a strain from the same serogroup and inoculation at the same dose and age), the first deaths classically occur 3 to 4 days after inoculation, and a high rate of mortality is reached in only a few days (Auldist et al., 1998; Bush & Staley, 1980; Hancock et al., 2002; Lacy-Hulbert et al., 1999). This corresponds to the pattern of evolution that is observed in the group W21 rabbits infected at 28 days. Thus, rabbits weaned early do not exhibit exacerbated sensitivity to this pathogenic strain compared to rabbits weaned just before the inoculation. It should also be noted that suckling rabbits presented an atypical pattern of evolution of colibacillosis. They were protected as long as they were suckling, but disease (clinical signs, mortality) started to develop as soon as they were weaned. Although the evolution of colibacillosis was less brutal and intense, it was quite similar to that in rabbits weaned earlier; however, it had a time lag compared to the rabbits weaned earlier. These observations suggest that milk plays an essential role against the virulence expression of EPEC. Several mechanisms of action, probably combined, may thereby be suspected. First, the *in vivo* growth of bacteria may have been stopped or reduced. Some milk components could directly exert an antimicrobial effect against pathogenic *E. coli* or could indirectly create unfavorable environmental conditions for their growth by modifying the commensal flora, the pH, or bacterial substrates. This hypothesis seems likely, as suckling rabbits presented a lower level of excretion of pathogenic *E. coli* in their feces until weaning. Moreover, according to the *in vitro* tests conducted, it seems that a direct bacteriostatic effect of some milk substances is implicated. Second, milk could protect against diarrhea by limiting bacterial adhesion to enterocytes. Indeed, the pathogenesis of EPEC infection starts by a first step of loose adhesion between bacteria and enterocytes mediated by specific adhesins (Farr et al., 2002; Griesbeck-Zilch et al., 2008). This hypothesis was confirmed by *ex vivo* results that showed a reduction in the level of adhesion of E22 bacteria to the ileum of suckling rabbits compared with that in weaned pups. It was also consistent with results obtained by using a recently developed assay of inhibition of AF/R2-dependent adhesion of EPEC to HeLa cells (Barrington et al., 1997). This assay is based on the ability of pathogenic *E. coli* O103 to adhere to HeLa cells *in vitro*, *via* the AF/R2 adhesin (Moore et al., 2005). In this study, inhibition of adhesion of *E. coli* to HeLa cells was demonstrated upon pre-incubation of the bacteria with milk and lactoserum as well as after pre-incubation of milk and lactoserum with HeLa cells before incubation of the bacteria with epithelial cells. These reports also suggest that some milk components may prevent adhesion of the bacteria to enterocytes by acting on fixation sites from both the bacterial and the epithelial cell sides. Indeed, the first step of epithelial colonization by *E. coli* strains results from the connection between bacterial pili and surface receptors on host cells. For instance, sialoglycoproteins act as receptors for the AF/R1 pilus of the rabbit RDEC-1 *E. coli* strain (Palmer et al., 2006), and glycosphingolipid and transferrin act as receptors for the adhesin K88 from some pig enterotoxigenic *E. coli* ETEC strains (Hooper et al., 2003). The exact nature of the receptor specific for AF/R2 has not been identified yet. Some minor components present in milk could interfere with the mechanisms leading to diarrhea, independent of bacterial colonization. Indeed, the strong inflammatory response induced by EPEC adhesion to the epithelium is at least partly responsible for diarrhea (Palmer, 1973; Mayer et al., 2005; Newburg, 2005). Several studies have also revealed the anti-inflammatory properties of human milk, which contains regulatory cytokines such as interleukin-10 and transforming growth factor β. Indeed, these molecules have been implicated, for instance, in the protective role of milk in infant necrotizing colitis (Fonteh et al., 2002; Horton, 1995). In addition, some healing properties of human milk have also been described, mainly in association with the presence of epidermal growth factor (Fonteh et al., 2002; Leitner et al, 2008; Ormrod & Miller, 1992). Interestingly, it has been shown that oral complementation of young rabbits with epidermal growth factor prevents the appearance of REPEC-induced diarrhea (Barrington et al, 2001).

The natural virtues of milk in protecting young mammals against numerous pathogenic viruses and bacteria have been known for many years (Petitclerc et al, 2007; Sanchez et al, 1992). A specific passive protection by maternal antibodies is often responsible for this phenomenon. However, in this study, the transmission of specific antibodies by maternal milk was not involved, since all other antibodies were seronegative for the O103 LPS. More recently, other substances present in milk have also been shown to have antimicrobial properties: defensins (Hodgkinson & Hodgkinson, 2003), lactoferrin (Rincheval-Arnold et al, 2002) and its derivative, the lactoferricin (Gopal & Gill, 2000), various enzymatic complexes like lactoperoxidase, xanthine oxidoreductase, and lysozyme (Mitra et al, 1995; Oviedo-Boyso et al, 2007; Phalipon et al, 2002), fucosylated oligosaccharides (Cheng et al,2008), and various fatty acids (Hodgkinson et al, 2007; Miles, 2004; Rainard & Riollet, 2006). It has been suggested that these components could inactivate pathogens by diverse mechanisms (growth inhibition, perturbation of adhesion) and may thereby act additively and/or synergistically (Hodgkinson et al, 2007). In humans, for instance, fucosylated oligosaccharides and lactoferrin have been shown to inhibit the localized adhesion of EPEC to HEp-2 and HeLa cells *in vitro* (Cheng et al, 2008; Christofferson, 2006). Thus far, the presence of these different components in doe's milk has not been demonstrated, although it has been shown that doe's milk is free from lactoferrin (Mookherjee & Hancock, 2007). It would, thus, be interesting to characterize the specific antimicrobial molecules present in doe's milk.

Management of Anti-Microbial Proteins

Assay of Lactoperoxidase Activity

Peroxidases catalyze reactions in which hydrogen peroxide is reduced and a suitable electron donor is subsequently oxidized. A wide variety of organic and inorganic substances can serve as electron donors, but substrate specificity varies between various peroxidases. Most assays are based on the principle that the electron donor in the oxidized form absorbs light and can be determined by a spectrophotometric assay. Currently 2, 2'-azino-di-3-ethylbenzthiazoline-6-sulphonic acid (ABTS) is accepted as the electron donor of choice for peroxidase assay. ABTS is relatively stable in aqueous solution and there is no information, to the best of this author's knowledge, to indicate that it is hazardous to health. The peroxidase assay using ABTS was originally developed by Shindler for the determination of peroxidase activity in cervical mucus. However, work by Björck and others has shown that the method is suitable for determining LP activity in milk and milk products. LP is a basic protein and, therefore, it can absorb to charged surfaces, particularly glass. However, it retains its catalytic activity when absorbed to glass and this can cause difficulties in assaying the enzyme. Glass cuvettes must be rinsed with acid between samples to remove any residual LP. Alternatively, disposable cuvettes can be used. Pruitt and his colleagues have also demonstrated that high concentrations of SCN (>0.5 mM) will interfere, giving lower LP activities due to simple competitive interference. While it is unlikely that levels >0.5 mM will be encountered in milk, it is advisable to measure SCN levels in unknown milk samples prior to critical LP analysis. Small differences in D A413/min values will have a significant effect on the results; for example, a difference of 0.1 in DD A413/min will result in a 10% difference in activity (U/ml) of a solution of LP.

Determination of Biological Activity or Antimicrobial Activity

Antimicrobial activity can be assessed by adding thiocyanate and hydrogen peroxide at appropriate concentrations to a suitable medium containing LP and a designated strain of *E. coli,* followed by determining the change in total viable count of the *E. coli* population. Alternatively, the effect on acid production by the *E. coli* can be measured. The latter method is limited in that it cannot provide information on the level of bactericidal effect, but it is simple and convenient to perform. The relationship between total enzyme-protein concentration and antimicrobial activity can be obtained by diluting an enzyme preparation of known concentration in a quantitative manner and determining the concentration at which LP system activation is limiting. Methods were originally developed to enable the *in vitro* evaluation of the antimicrobial activity of calf milk replacers containing an LP system. There is no standard reference method, and the evolution of the approach has been described. Skimmed milk prepared from medium-heat skimmed milk powder, free from antibiotic residues and inhibitors of the LP system and containing no detectable LP activity, is reconstituted using distilled water to give 10% (w/v) milk solids. A simple acidification assay can be performed using *E. coli* NCDO 904. Tubes of milk (10 ml), inoculated with *E. coli* NCDO 904 to give 107 CFU/ml and supplemented with thiocyanate (0.25 mM) and hydrogen peroxide (0.25 mM), are incubated at 37°C for 6 h. pH values at 6 h are subtracted from time 0 values and compared with controls containing 5 μg LP/ml and ± L- cysteine at an added concentration of 1 mM. The control without cysteine should give little or no change in pH, whereas tubes containing 1 mM cysteine should show pH values ranging from 0.25 to 0.4. *E. coli* NCDO 2328 is much more sensitive to the LP system than strain NCDO 904. Consequently, the bactericidal effects of the LP system are studied most easily using NCDO 2328. Media based on milk, as discussed previously, or 0.01 M phosphate buffer can be used. The method is essentially as described for the acidification assay, except that viable calls of *E. coli* are enumerated using Violet Red Bile Agar (VRBA) at time 0 and after 2 h incubation. Antimicrobial effects can be determined in 0.01 M phosphate buffer using commercial LP preparations diluted to give LP activity equivalent to that given by enzyme concentrations between 0.05 and 0.5 μg/ml. Optimal antimicrobial activity requires >0.5 μg LP/ml. Specific details on factors affecting precision and accuracy have not been published in full, but are known to include: strain of *E. coli* used, temperature of assay, incubation time and presence or absence of inhibitors of the LP system. Controls incorporating a specific inhibitor of the LP system, for example L-cysteine, are required.

Antibacterial and Antiviral Effects of Milk Proteins and Its Derivatives

Milk forms a rich source of biologically interesting components. In particular, its protein fraction is known to encompass many kinds of biological functions and include chemically-modified proteins and enzymatically-induced peptides. If such peptides are released by enzymes present within the digestive tract (e.g., trypsin or pepsin), it is likely that they play a role in the health defense system. This is especially the case when active fragments can survive the intestinal conditions long enough to arrive at the right place to exert their beneficial function. In the earlier part of this chapter, the antibacterial proteins lactoferrin, lactoperoxidase, and lysozyme have been described in detail. Furthermore, antibacterial peptides originating from caseins and whey proteins are described. The second part reports on studies of antiviral effects of milk proteins and derivatives thereof. Special focus is directed to the antiviral action towards the human immunodeficiency virus (HIV) and the human cytomegalovirus (HCMV). Unmodified milk proteins are generally not active against these viruses. An exception is lactoferrin, which shows significant antiviral activity against both HIV and HCMV. Several other milk proteins tested have showed strong antiviral effects only after chemical modification, i.e., by making them polyanionic (for anti-HIV activity) or polycationic (for anti-HCMV activity). In a number of cases, conclusions are drawn concerning possible relationships between antibacterial/antiviral activity and molecular structure of the components described.

Lactoferricin

Lactoferricin (LFcin) was initially identified as an antimicrobial peptide derived by pepsin digestion of lactoferrin (LF), a multifunctional innate-defense protein in milk. Various synthetic analogs of LFcin have also been reported. LFcin inhibits a diverse range of microorganisms such as Gram-negative bacteria, Gram-positive bacteria, yeast, filamentous fungi, and parasitic protozoa, including some antibiotic-resistant pathogens. LFcin kills target organisms by membrane perturbation and acts synergistically with some antimicrobial agents. LFcin exhibits numerous biological activities in common with those of LF. While LFcin suppresses the activation of innate immunity by microbial components such as lipopolysaccharide (LPS) and CpG DNA, the peptide itself activates immunity. Administration of LFcin analogs has been shown to protect the host *via* direct antimicrobial activity and immuno-stimulatory effects in several infection models of mice. Here, we present a comprehensive review of investigations of LFcin and related peptides.

The Antibacterial Properties of Breast Milk

Many years ago, breast milk was thought to be sterile. While this is far from being the case (it actually contains all manner of germs), the role it plays in helping keep babies free from harmful diseases means it does display impressive immunogenic capacities. The immunological components of breast milk help to protect both a mother's breast and her baby from infection during feeding, as well as aiding the development of the baby's immune system. They also have another useful consequence, that of protecting breast milk from disease for some time after it has been expressed, enabling it to be stored. Several studies have examined whether it is safe to keep expressed milk for short periods, and there is general agreement that it can be stored for 8 hours at room temperature (25 degrees C), for three days in the fridge (4 degrees C) and for up to a year in the freezer (-20 degrees C) without any increase in the levels of pathogens. Breast milk not only inhibits the growth of pathogens but it actively reduces them. This was convincingly demonstrated in a research study examining what happened to milk during short-term storage. Milk was collected from 9 mothers and divided into three samples: the first was analyzed the same day; the second was refrigerated (at 4 to 6 degrees C) for 48 hours; and the third was refrigerated for 72 hours. Each sample was then contaminated with an *E.coli* suspension, and left for two hours. When the samples were tested, levels of *E.coli* had reduced by 80% in both the milk that was fresh and the milk that was 48 hours old.

Levels had also diminished in the 3 day-old milk, but only by around 10%, indicating that the antibacterial properties, whilst still present, had started to degrade by this point.

If a baby needs to feed from a bottle or cup, a considerable body of research indicates that it is safe to give him/her breast milk that has been stored in the fridge for up to three days, or in the freezer for several months. There is also evidence that if the expressed milk does come into contact with germs, then the bactericidal components of breast milk should be able to take care of them, providing the milk is under two days old. There may still be potential issues associated with feeding stored, rather than fresh breast milk to the baby: various chemical changes occur in milk once it has left the body, and it is possible some of these may affect its nutritional value. Nevertheless, expressed breast milk remains a healthier alternative than formula milk, and as a result of its antibacterial qualities, one can rest assured of its positive impact on the baby's health.

Xanthine Oxidase

Though the abundance of xanthine oxidase (XO) in milk has been recognized for more than a century (Hodgkinson & Hodgkinson, 2003), its physiological function has remained a matter of speculation. Isolated studies, however, have considered an antimicrobial role for XO; in these, purified XO was shown to inhibit bacterial growth in the presence of a reducing substrate, which presumed the generation of hydrogen peroxide (Farr et al., 2002; Hodgkinson, et al., 2007; Leitner et al., 2008). Recently, the ability of XO to catalyze reactions that generate nitric oxide ($NO^{\cdot}$) from inorganic nitrite was reported (Christofferson, 2006; Hooper et al., 2003). $NO^{\cdot}$ is formed only under anaerobic conditions; when oxygen is present, superoxide is also produced, which reacts with XO-generated $NO^{\cdot}$ to form peroxynitrite (Clark et al., 2006), a powerful bactericidal agent (Barrington et al., 2001). These observations prompted us to investigate the antibacterial activities of milk, particularly under defined oxygen tensions and with reference to the involvement of nitrite. *Escherichia coli* cells transformed with the plasmid pLITE27 that carries the luxCDABE operon from *Photorhabdus luminescens* controlled by a constitutive promoter were used. Microorganisms expressing the lux operon are able to emit a blue-green light due to the activity of bacterial luciferase on reduced flavin mononucleotide and a long-chain aldehyde that produces flavin mononucleotide since reduced flavin mononucleotide production depends upon functional electron transport, only metabolically active cells can produce light (Auldist et al., 1998). Therefore, lux operon-dependent bioluminescence is an extremely sensitive, nondestructive, real-time reporter of cell metabolism that has been successfully used to monitor antimicrobial effects (Lacy-Hulbert et al., 1999). This technology confers many advantages over conventional plating techniques and has allowed the first-time demonstration of bacteriostatic activity for both bovine and human milk that is dependent on XO, low oxygen tensions, and nitrite. *E. coli* isolate 16906 was transformed with the luxCDABE operon (Lacy-Hulbert et al., 1999). Cells were grown overnight at 37°C on nutrient agar plates with 50 μg of ampicillin per ml, sub-cultured into minimal salts medium (Bush & Staley, 1980), and grown at 37°C with aeration for 3 h. Aliquots were mixed with an equal volume of 40% glycerol and frozen at −80°C. Before each assay, aliquots were pelleted by centrifugation and re-suspended in fresh minimal salts medium at 37°C without glucose or a nitrogen source. Suspension of the cells in the nutrient-deprived medium caused the cells to cease active growth and enter stationary phase, where metabolic activity is at a steady low level. Measured light output was stable, as monitored by a specifically adapted luminometer. This instrument (LKB 1250) was modified so that the reaction cuvette could be continuously supplied with a defined mixture of air and nitrogen, thus allowing accurate control of oxygen tension, which was measured with a Clark-type oxygen electrode. Additions of substrate or inhibitors were made through additional injection ports, effecting minimal disturbance of either oxygen tension or signal output.

The effects of bovine or human milk on the metabolic activity of *E. coli* were studied under conditions of 0.63% oxygen. This oxygen concentration is known to be optimal for the generation of peroxynitrite by XO (Clark et al., 2006). In these experiments, the rapid fall and rise in light emission immediately following the addition of milk were deemed to be artefactual, given that they occur with boiled milk. While clearly independent of XO activity, these rapid changes probably reflect the quenching effects of milk on the luminescence signal. However, the second large fall in luminescence output was not observed when heat-treated milk was used. Furthermore, the effect was eliminated by the addition of oxypurinol, a specific inhibitor of XO; therefore, this activity could be attributed to this enzyme. This was the first real-time measurement of the effect of milk on bacterial metabolic activity and the first demonstration of the nitrite dependence of this effect. Previous studies have been concerned primarily with purified XO (Farr et al. 2002; Hodgkinson et al., 2007; Leitner et al., 2008), although the results of two of these (Hodgkinson et al., 2007; Leitner et al., 2008) indicated that addition of the reducing substrate, hypoxanthine, led to decreased bacterial growth in samples of whole milk, and that the antibacterial effects of purified XO were attributable to hydrogen peroxide. The antibacterial role in the neonatal gut for XO-generated $NO^•$, and potentially peroxynitrite, is credible for a number of reasons, both spatial and catalytic. XO is located on the outer surface of the milk fat globule membrane, and pathogenic bacteria, with the capacity to target epithelial membranes of the digestive tract, may well bind to similar antigens on the milk fat globule membrane, which is itself of epithelial cell origin (Griesbeck-Zilch et al., 2008, Horton, 1995). This process not only diverts the bacteria from their primary target but also brings them into intimate contact with XO. Contact of this type will be further promoted by the known affinity of XO for acidic polysaccharides (Akira & Takeda, 2004), which occur in many bacterial capsules (Gopal & Gill, 2015). Therefore, the local concentrations of $NO^•$ and peroxynitrite produced in the vicinity of the bacteria could potentially be very high and could account for the antibacterial effects observed.

The catalytic properties of XO are also well suited to its proposed antibacterial function. The optimal pH for anaerobic XO-catalyzed generation of $NO^{\cdot}$ is pH 6 or less (Christofferson, 2006), much lower than pH 8.8 (which is optimal for aerobic XO activity) and more suited to a role in the digestive tract. It has been suggested that the pH of the neonatal gut generally ranges between pH 4 and pH 6 (Leitner et al., 2008), while others have suggested that the gut of a newborn immediately following birth is relatively alkaline, becoming progressively acidic during the first few days of life (Larson et al., 1980). Oxygen tensions are also low in the neonatal gut, and therefore, the conditions used in the present study are both representative of physiological conditions suggested by others and ideal for $NO^{\cdot}$ generation by XO (Leitner et al., 2008). Moreover, while the Km value for nitrite in XO-dependent $NO^{\cdot}$ production is in the millimolar region (Christofferson, 2006), this level can be achieved by enteric bacteria. In anaerobic culture, such bacteria can excrete millimolar levels of nitrite (Butler 1973; Komine et al., 2006) derived from dissimilatory nitrate reductase (Boots & Floris. 2006). It is an intriguing thought that, by generating a nitrite-rich microenvironment, enteric bacteria might initiate their own destruction. Finally, it is worth noting that XO activity of human milk, while generally very much lower than that of cows' milk (Adamski & Demmer, 2000), is exceptionally high in the first few weeks postpartum (Barrington et al., 1997). This is precisely the period when antibacterial activity is required in the neonatal gut and when it both coincides with particularly high levels of nitrite (Fonteh et al., 2002) and correlates with the highest levels of XO-dependent $NO^{\cdot}$ generation found in human milk (Leitner et al., 2008). In summary, it was demonstrated by Leitner et al., (2008) that the antibacterial activity of bovine and human milk is dependent on XO-catalyzed reduction of nitrite, leading presumably to $NO^{\cdot}$ and, eventually, peroxynitrite. These findings not only contribute significantly to the long-standing question of the physiological significance of XO in milk but are also relevant to the debate concerning the relative merits of breast- and bottle-feeding of infants. While it is well established that formula-fed infants in developing countries suffer from higher levels of gastrointestinal infections than do breast-fed infants (Apodaca et al., 1994; Cheng et al., 2008; Hancock et al., 2002), the reasons for this are not clear. A possible factor is XO activity, which is known to be lacking in formula feed (Leitner et al., 2008). Human and bovine milk inhibited the metabolic activity of *Escherichia coli*, as shown by luminescence monitoring of constructs expressing the luxCDABE genes. Inhibition was dependent on both xanthine oxidase (XO) activity and on the presence of nitrite, implying that XO-generated nitric oxide functions as an antibacterial agent.

Antimicrobial Activity of XOR in Milk

A third contribution to antimicrobial defense occurs in the special case of the neonatal gut. As noted above, the antibacterial activity of purified bovine milk XOR has long been recognized. Green and Pauli (Mitra et al., 1995) and Lipmann and Owen

(Rincheval-Arnold et al., 2002) demonstrated the growth inhibition of *Staphylococcus aureus* by purified bovine milk XOR and explained their findings in terms of the cytotoxic properties of XOR-generated hydrogen peroxide. Similar antibacterial activity has been confirmed in the cases of *Escherichia coli* and *Salmonella enteritidis* (Willix-Payne et al., 2001), while Bjorck and Claesson (Boots & Floris. 2006) proposed that XOR-derived hydrogen peroxide exerts its antimicrobial effects by acting as a substrate for the lactoperoxidase system in milk. In view of the recently discovered capacity of XOR to reduce nitrite to NO (see above), generation of reactive nitrogen species (RNS), particularly peroxynitrite, can be seen as a potential mechanism of antimicrobial activity of milk. As discussed above, nitrite concentrations are likely to be high in the immediate microenvironment of enteric bacteria, favoring XOR-catalyzed reduction to NO. Moreover, several factors promote the physical association of such bacteria with XOR, which is present on the Milk Fat Globule Membrane (MFGM). The MFGM is derived from the apical membrane of the mammary secretory cell and, as such, bears epithelial membrane antigens. Pathogenic bacteria target epithelial membranes of the digestive tract and may well bind to similar antigens on the MFGM (Oviedo-Boyso et al., 2007). This not only diverts the bacteria from their primary target but also brings them into intimate contact with XOR. Such contact is further promoted by the known affinity of XOR for acidic polysaccharides, such as those that occur in many bacterial capsules (Akira & Takeda, 2004).

In the generation of peroxynitrite, oxygen and nitrite compete for XOR-generated electrons, and there is an optimal oxygen tension for peroxynitrite production (Leitner et al., 2008). In the case of the oxidase form of the enzyme, XO, this tension is low, at around 8 μM. With XDH, on the other hand, relatively high rates are maintained over the physiological range of oxygen tensions. Moreover, the maximal rate of peroxynitrite generation is approximately double the maximal rate for XO generation. It is noteworthy that it is the XDH form of the enzyme that predominates in freshly expressed milk (Sordillo et al., 1997).

In support of the rationale described above, experimental evidence was obtained for the antibacterial activity of milk that appears to depend on XOR-generated RNS. Samples of fresh human milk and bovine milk were each shown to inhibit the metabolic activity of *E. coli*. A transformed *E. coli* construct that constitutively expressed the luxCDABE operon was used (Vorbach, et al., 2006). Impairment of the metabolic function results in reduced light emission, which can be quantitatively and continuously monitored. The milk-induced metabolic inhibition of *E. coli* was blocked by oxypurinol, a specific inhibitor of XOR (Moore et al., 2005). It was also found to be dependent on the presence of nitrite (Moore et al., 2005) and to be optimal at low oxygen tensions when XOR production of NO is maximal (Leitner et al. 2008).

Role of Xanthine Oxidoreductase as an Antimicrobial Agent

- Milk contains several antimicrobial components, including lactoferrin, lactoperoxidase, lysozyme, and possibly N-acetyl-ß-D-glucosaminidase, which may be involved in protecting against mastitis, protecting against bacterial growth post-harvest, and protecting the consumer of the milk product.
- Milks of different species contain different amounts of the various antimicrobial factors. Cow milk has high lactoperoxidase, but low lactoferrin and lysozyme, while human breast milk has high lactoferrin and lysozyme, but low lactoperoxidase.
- The ability to alter the activity of these anti-microbial factors in cow milk could have an impact on shelf-life of raw milk and development of additional health and functional foods based upon these factors.

N-Acetyl-ß-d-Glucosaminidase

N-Acetyl-ß-D-glucosamindase (NAGase) is an enzyme whose activity has been implicated as an indicator of tissue damage during mastitis. It is a lysosomal enzyme that is secreted in large quantities in the mammary gland during involution and inflammation. The NAGase enzyme has also been found in other bovine secretions, such as uterine fluids. Though the specific function of NAGase in the mammary gland is not known, recent research has suggested that NAGase may exhibit some antimicrobial activity. During lactation, cow milk normally has low NAGase activity. Similarly, NAGase is low in mammary secretions in the early dry period, coinciding with the period of highest incidence of new intra-mammary infection. By the mid-dry period, however, NAGase activity is at its highest in mammary secretions, concurrent with the lowest incidence of new intra-mammary infection. Therefore, high levels of NAGase activity, along with elevated lactoferrin concentrations in the mammary gland during the mid-dry period may contribute to increased antibacterial activity found in mammary secretions at that time. There is a relationship between the presence of pathogens in the udder and NAGase levels in milk. Marked increases in NAGase activity resulting from the presence of major mastitis pathogens have been observed. Since NAGase has been found in uterine fluids, it has been suggested that NAGase may have a role in the bactericidal function of the uterus as well. Researchers have studied the bactericidal effect of NAGase on several bacterial pathogens commonly found to infect the cow uterus. Of these pathogens, *Actionmyces pyogenes*, *Staphylococcus aureus*, *Strepto-coccus agalactiae*, and *Pseudomonas aeroginosa* were inhibited by NAGase, while the *Escherichia coli* and *Enterobacter aerogenes* were not inhibited. Although these results cannot be directly extrapolated to bacterial strains that cause mastitis, they do lend support for such an antimicrobial function of NAGase in the mammary gland.

Colostrum as Source of Antimicrobial Agents

Colostrum and milk provide a complete diet for the neonate. In ruminants, colostrum is also the sole source of initial acquired immunity for the offspring. Milk, therefore, plays an important role in mammalian host defense. In colostrum, the concentration of immunoglobulins is particularly high, with IgG being the major immunoglobulin class present in ruminant milk, in contrast to IgA being the major immunoglobulin present in human milk. Immunoglobulins are transported into mammary secretions via specialized receptors. In addition to immunoglobulins, both colostrum and milk contain viable cells, including neutrophils and macrophages, which secrete a range of immune-related components into milk. These include cytokines and antimicrobial proteins and peptides, such as lactoferrin, defensins, and cathelicidins. Mammary epithelial cells also contribute to the host defense by secreting a range of innate immune effector molecules. A detailed understanding of these proteins and peptides offers great potential to add value to the dairy industry. The immune function of milk, in particular, how the gland responds to pathogens, can be used to boost the concentrations of immune factors in milk through farm management practices and vaccination protocols. This approach is currently being used to maximize yields of bovine milk-derived IgA directed at specific antigens for therapeutic and prophylactic use. Increasingly sophisticated proteomic-based technologies are being applied to identify and characterize the functions of the minor components of milk. In ruminants in particular, for which no exchange of immune factors occurs *in utero*, colostrum and, to a lesser extent, milk provide protection through a high immunoglobulin content, without which the ruminant would not survive. The colostrum immunoglobulins, in conjunction with the ability of the ruminant neonatal gut to allow unrestricted passage of the large immunoglobulin molecules, provide the young animal with passive immunization. However, both the concentration of immunoglobulins in colostrum and the permeability of the gut decrease rapidly and progressively over the first 48 h after birth. Therefore, an adequate supply of colostrum, with abundant immunoglobulins, is essential during this brief period of time for the young to gain sufficient passive immunity to be able to survive until its own immune system is fully developed. Immune factors in colostrum and milk also play an important role in the host defense of the mammary gland itself, protecting it from pathogenic organisms. The mammary gland is very susceptible to infection because mammary secretions provide an excellent source of nutrients for pathogens, the body temperature is ideal for bacteria growth, and the gland, *via* the teat opening, has a direct link to the external environment. The mammary host defense system has evolved into an extremely well-developed, complex, and highly effective barrier against pathogens, integrating both the innate and acquired immune response. A wide variety of components linked to the innate immune response have been identified in colostrum and milk. These include neutrophils, macrophages, complement, oligosaccharides, gangliosides, reactive oxygen species, acute-phase proteins, immunomodulatory factors (including many different pro- and anti-inflammatory cytokines), ribonucleases (RNase), and a range of peptides and proteins with direct antimicrobial activity. Many of these components originate from specialized cells that traffic to the mammary gland. For example, neutrophils and macrophages infiltrating the mammary gland not only kill bacteria directly through phagocytosis but are also responsible for the production of many cytokines, reactive oxygen species, and antimicrobial peptides. Several recent reports surveying the minor constituents of milk from healthy and infected glands have noted the predominance of host defense-related proteins. However, it is beyond the scope of the current overview to discuss in detail the role and mechanisms of action of all the innate immune factors that have been described in colostrum and milk. The bovine mammary epithelium also plays an active role in host defense and synthesizes several innate immune factors, such as lactoferrin, β-defensin, and lipopolysaccharide-binding protein. Moreover, preparations of the membranes surrounding the milk fat globules secreted by MEC, which originate from the apical membrane of the MEC, contain many peptides, proteins, and lipids that are involved in host defense. Key milk fat globule membrane components, such as xanthine oxidase and sphingolipids, have antibacterial activity against a range of bacteria.

Applications of Immune and Antimicrobial Components to Add Value to Milk

Milk production by the modern dairy cow is far in excess of the nutrient requirements of its calf, and milk has become a significant economic commodity. Traditionally, milk has been used by humans for direct consumption or as a raw ingredient for manufacturing cheese, butter, fermented dairy products, or milk powder. However, recent proteomic studies of human colostrum and bovine milk have revealed a large number of minor components, many of which have an immune function. Such immune components hold great potential to add significant value to milk, with applications in pet food, cosmetics, personal care, and health promotion.

Innate Immune Antimicrobial and Immuno-Modulatory Factors

Small cationic peptides and proteins of the innate immune system are increasingly valued for their potential as antibacterial, antifungal, or antiviral products, or their combination They may even hold potential as natural alternatives to traditional antibiotics, because the development of resistance may be less of a problem with such innate molecules. Several antimicrobial molecules are currently at various stages of development in several biotechnology companies. Although such antimicrobials could be extracted from a wide range of biological sources, including plants, insects, and marine organisms, milk lends itself particularly well to mining such components. Milk can easily be harvested in liquid form, it is already processed on a large scale by dairy manufacturers, and, importantly, it contains many immune components. Milk-derived lactoferrin was one of the first immune components of milk to be commercially extracted for its antimicrobial and antiviral properties. It can be found as a valuable ingredient in infant formula and other foods for both human and pet consumption, in skin care products (e.g., cosmetics), and in oral care products such as toothpaste, mouthwash, and chewing gum. Lactoferrin, in combination with lactam antibiotics, also shows promise in the treatment of bovine mastitis. Lactoperoxidase, another antimicrobial protein commercially extracted from milk, is used for food preservation, oral care, and even as a plant fungicide. The lytic enzyme lysozyme, which is present in milk, also has application as a food preservative in the food industry. Despite the commercial applications of some milk-derived immune components to date, the commercial potential of the majority of immune-active components identified in milk remains untapped. Many of these components have activities that are not directly antimicrobial. Instead, they contribute to the generation of a robust and appropriate immune response through such mechanisms as the detection of pathogens, the recruitment of immune cells such as neutrophils, the promotion of phagocytosis and cell-mediated killing, activation of complement pathways, and controlled resolution of inflammation. The extraction of such components could provide a new class of milk-derived, immune-related bioactive compounds that prevent infection or promote the resolution of disease by directly modulating the host immune response. The development of such immuno-modulatory bioactive compounds does pose challenges. In particular, the host immune response is a fine balance between pro- and anti-inflammatory components, and manipulating this response requires care. However, these bioactive compounds do offer powerful strategies for preventing or treating diseases that are not dependent on antibiotics and the associated difficulties of antibiotic resistance. Given the range of soluble immune active components that have now been characterized in milk and their ease of extraction, it is likely that some of these proteins will form the basis for future products and treatments that directly modulate the host immune systems of humans or production animals.

Antimicrobial Characteristics of Milk Products

Probiotics are living, health-promoting microorganisms that are incorporated into various kinds of foods. The ability of probiotics to withstand the normal acidic conditions of the gastric juices and the bactericidal activity of the bile salts, as well as the production of lactic acid that inhibits the growth of other microorganisms, allow them to be established in the intestinal tract. The reported health benefits include: boosting of the immune system, inhibition of the growth of pathogenic organisms, prevention of diarrhea from various causes, prevention of cancer, reduction of the risk of inflammatory bowel movements, improvement of digestion of proteins and fats, synthesis of vitamins, and detoxification and protection from toxins. The concept of probiotics progressed around 1900, when Elie Metchnikoff hypothesized that the long and healthy lives of Bulgarian peasants were the outcome of their consumption of fermented milk and milk products. Members of the genera *Lactobacillus, Bifidobacterium* and *Streptococcus* are the most common probiotics used in commercial fermented and non-fermented dairy products today. Among the popular probiotic foods in the world market today are some infants' milk, fermented milk-based drinks and yogurts, probiotic curd, and probiotic cheese. Various studies were conducted to determine the presence of antimicrobial activities among the probiotics incorporated into these different food products against common microbial pathogens. Substantiating the antimicrobial activities of probiotics affirms their use in the development of functional foods for the betterment of the health of the consuming public. Probiotics refer to the viable mono or mixed cultures of microorganisms, which on ingestion in certain doses exert beneficial health effects to the host by improving the properties of indigenous microflora in the gastro-intestinal (GI) tract. Probiotics are mostly selected from the lactic acid bacteria (LAB) group in which *Lactobacillus* species normally predominate because of their capacity to grow rapidly and tolerate acidic environments. Probiotics bring about changes in enteric flora with decrease in *E.coli* count and colony formation on the intestinal wall, thereby preventing colonization by pathogenic microorganisms, synthesis of lactate with consequent decrease in intestinal pH, neutralization of toxic amines and ammonia, generation of antibacterial substances, immunomodulation, and competition with other microorganisms for nutrients in the GI tract. (Korhonen & Pihlanto, 2006). Probiotics put up ecological barriers against the infective bacteria. Production of H_2O_2 and bacteriocins by these organisms prevents proliferation of pathogens. They can prevent diarrhea. Probiotics exhibit anti-mutagenic and anti-carcinogenic activity, inhibit cholesterol synthesis by the body, stimulate the immune system against certain antigens and pathogens, reduce lactose intolerance by metabolizing lactose, improve calcium absorption, synthesize vitamins and help in pre-digestion of proteins (Schlimme & Meisel, 1995). LAB are extensively used in foods because of their role as preservatives, nutritive and therapeutic agents. They enhance longevity of humans and are natural inhabitants of the GI tract. Probiotics for application in value-added dairy foods may include lactobacilli such as *L. acidophilus, L. plantarum, L. casei subsp. rhamnosus GG, L. brevis, L. delbruckii subsp. bulgaricus* etc.; bifidobacteria such as *B. bifidum, B. infantis, B. longum, B. animal is, B. breve* etc.; yeast such as *Candida kefir, Kluveromyces fragilis, Saccharomyces boulardii* or others like *Enterococcus faecium, Lactococcm lactis subsp. lactis, Lactococcus lactis subsp. cremoris* etc. (Vermeirssen et al., 2003a).

Antimicrobial Bio-Peptides from Cheese/Casein

There is no doubt that milk proteins provide excellent nutrition for the suckling. However, apart from that, milk proteins can also provide numerous physiological activities benefiting the suckling in a variety of ways. These activities include enhancement of immune function, defense against pathogenic bacteria, viruses, and yeasts, and development of the gut and its functions. Besides the naturally occurring biologically active proteins present in milk, a variety of bioactive peptides are encrypted within the sequence of milk proteins that are released upon suitable hydrolysis of the precursor protein. A large range of bioactivities has been reported for milk protein components, with some showing more than one kind of biological activity (Schlimme & Meisel, 1995). This chapter reviews the most important antimicrobial peptides derived from milk proteins, especially those that may have a physiological significance to the suckling neonate and also exhibit a range of biological activities that influence digestion, metabolic responses to absorbed nutrients, growth and development of specific organs, and resistance to disease. These biological activities are mainly due to the peptides and proteins in milk. However, some of the biological activity of milk protein components is latent, and is released only upon proteolytic action. Bioactive peptides are produced during digestion of milk in the gastrointestinal tract, and also during fermentation and food processing.

Bioactive peptides have been defined as specific protein fragments that have a positive impact on body functions or conditions and may ultimately influence health. Upon oral administration, bioactive peptides may affect the major body systems, namely, the cardiovascular, digestive, immune and nervous systems. The beneficial health effects may be classified as antimicrobial, antioxidative, antithrombotic, antihypertensive, antimicrobial or immuno-modulatory. Antimicrobial peptides (also called host defense peptides) are an evolutionarily conserved component of the innate immune response and are found among all classes of life. Antimicrobial peptides have captured the attention of researchers in recent years because of their efficiency in fighting against pathogens. These peptides are found in nature and have been isolated from a wide range of organisms. Furthermore, analogs or synthetic derivatives have successfully been developed on the basis of natural peptide patterns. Long use of pesticides and antibiotics has led to the development of resistance among pathogens and other pests as well as increased environmental and health risks. Antimicrobial peptides are under consideration as new substitutes for conventional pesticides and antibiotics. Many plants and animals have been manipulated with antimicrobial peptide-encoding genes and several pesticides and drugs have been produced based on these peptides. Such strategies and products may still have a long way to go before being confirmed by regulatory bodies, and need to surmount technical problems before being accepted as applicable ones. In spite of these facts, several cases of successful use of antimicrobial peptides in agriculture and food industry indicate a promising future for extensive application of these peptides. Antimicrobial peptides have been demonstrated to kill Gram negative and Gram positive bacteria (including strains that are resistant to conventional antibiotics), mycobacteria (including *Mycobacterium tuberculae*), enveloped viruses, fungi and even transformed or cancerous cells. Unlike the majority of conventional antibiotics it appears as though antimicrobial peptides may also have the ability to enhance immunity by functioning as immuno-modulators. Antimicrobial peptides are a unique and diverse group of molecules, which are divided into subgroups on the basis of their amino acid composition and structure. Antimicrobial peptides are generally 12-50 amino acid long. These peptides include two or more positively charged residues provided by arginine, lysine or, in acidic environments, histidine, and a large proportion (generally >50%) of hydrophobic residues. The secondary structures of these molecules follow 4 themes, including i) α-helical, ii) β-stranded due to the presence of 2 or more disulfide bonds, iii) β-hairpin or loop due to the presence of a single disulfide bond and/or cyclization of the peptide chain, and iv) extended. Many of these peptides are unstructured in free solution, and fold into their final configuration upon partitioning into biological membranes.

Production of Bioactive Peptides

Bioactive peptides are inactive within the sequence of the parent protein and can be released in three ways: (a) enzymatic hydrolysis by digestive enzymes, (b) food processing and (c) proteolysis by enzymes derived from microorganisms or plants.

a) Enzymatic Hydrolysis

The cleavage of latent bioactive peptides from milk proteins normally occurs during digestion by pepsin and pancreatic enzymes (trypsin, chymotrypsin, carboxy and aminopeptidases), producing active peptide fragments in the gastrointestinal tract of the milk-consuming individual (Chandan, 2000). The physiological effects of bioactive peptides depend on their ability to reach their target sites intact, which may involve absorption through the intestinal epithelium prior to traveling to the peripheral organs (Mishra & Prasad 2000). In recent years, research on the production of active peptides obtained from milk and their potential functionality has grown to a great extent. Bioactive peptides have been defined as specific protein fragments that have a positive impact on body functions or conditions, and they may ultimately have an influence on health. Individual proteins of casein or milk-derived products such as cheese and yogurt have been used as a protein source to study the isolation and activity of peptides with several applications. Currently, the milk whey waste obtained in the production of cheese also represents a protein source from which active peptides could be isolated with potential industrial applications. The active properties of milk peptides and the results obtained with regard to their physiological effects have led to the classification of peptides as belonging to the group of ingredients of protein nature, appropriate for use in functional foods or pharmaceutical formulations. Many of the known bioactive peptides have been produced *in vitro*, using gastrointestinal enzymes, usually pepsin and trypsin. ACE-inhibitory peptides and CPPs, for example, are most commonly produced by trypsin. Other digestive enzymes and different enzyme combinations of proteinases, including alcalase, chymotrypsin, pancreatin, pepsin and thermolysin as well as enzymes from bacterial and fungal sources, have also been utilized to generate bioactive peptides from various proteins.

b) Dairy and Food Processing

The structural and chemical changes that occur during the processing of food proteins may result in the release of bioactive peptides. In particular, heat and/or alkali treatment can generate additional inter-and intramolecular covalent bonds that are resistant to hydrolysis. Such processing conditions also promote the racemic conversion of L-amino acids to D-isomers and consequently, lead to indigestible peptide bonds. The potential formation of indigestible peptide sequences during food processing is noteworthy, because this may promote both formation and absorption of bioactive peptides that do not occur naturally in the precursor protein. Such bioactive peptides can be generated during manufacture of several milk products and may thus be ingested as food components. For example, partially hydrolyzed milk proteins for hypoallergenic infant formulae and for clinical applications in nutrition consist exclusively of peptides and contain bioactive peptides. Cheese contains phospho-peptides as natural constituents and secondary proteolysis during cheese ripening leads to formation of various ACE inhibitory peptides.

c) Microbial Fermentation

Many industrially utilized dairy starter cultures are proteolytic to some extent. Bioactive peptides can, thus, be generated by the proteolytic activities of the strains of starter and non-starter bacteria e.g., *Lactobacillus helveticus, Lactobacillus delbrueckii ssp. bulgaricus, Lactobacillus plantarum, Lactobacillus rhamnosus, Lactobacillus acidophilus, Lactococcus lactis*, and *Streptococcus thermophilus* used in the manufacture of fermented dairy products. The proteolytic system of lactic acid bacteria (LAB) is well-characterized. This system consists of a cell wall-bound proteinase and a number of distinct intracellular peptidases, including endopeptidases, aminopeptidases, tripeptidases and dipeptidases. Extracellular proteinases cause degradation of casein into oligopeptides. The longer chain oligopeptides may be a source of bioactive peptides when further degraded by intracellular peptidases of lysed-lactic acid bacteria.

The single most effective way to increase the concentration of bioactive peptides in fermented dairy products is to ferment or co-ferment with highly proteolytic strains of LAB. The choice of strains influences the release of effective bioactive peptides. The strain should not be too proteolytic, otherwise the product will be destroyed and must have the right specificity to give high concentrations of active peptides. The concentration of ACE-inhibitory peptides seems to rely on a balance between their formation and further breakdown into inactive peptides and amino acids that, in turn, depends on storage time and conditions. Various bioactive peptides including ACE-inhibitory or antihypertensive peptides, immuno-modulatory, antioxidative, and antimutagenic peptides have been released from milk proteins through microbial proteolysis (Mishra & Prasad 2000). The best known ACE-inhibitory peptides, Val-Pro-Pro (VPP) and Ile-Pro-Pro (IPP), have been identified in milk fermented with strains of *Lb. helveticus* and *Saccharomyces cerevisiae.* In addition to live microorganisms, proteolytic enzymes isolated from LAB have been successfully employed to release bioactive peptides from milk proteins.

CONCLUSION

Antimicrobial substances have important applications in the dairy, animal production, and human health industries. Based on the physiological and functional properties attributed to lactoferrin, a number of interesting and innovative possibilities for the application of this natural protective protein can be considered. Health and functional foods, sports nutrition capsules, and drinks containing lactoferrin are being developed. Lactoferrin is already being used as an ingredient in infant formulas. The antibacterial properties of the lactoperoxidase system have also been applied to animal production and clinical medicine. For example, due to the lack of cooling equipment, many farmers in China are faced with the problem of milk spoilage during storage and transport. In order to preserve the quality of milk, Chinese scientists are teaching farmers how to activate the lactoperoxidase system in raw milk. Addition of a small quantity of sodium thiocyanate and sodium percarbonate to fresh raw milk is effective at reducing milk spoilage. Another application could be the addition of lactoperoxidase to calf milk replacers as a substitute for antibiotics. Improved performance of calves by the activated lacto-peroxidase system has been demonstrated. Lysozyme's ability to limit the migration of neutrophils to damaged tissue means that it might be used as an anti-inflammatory agent. Lysozyme could be used as an additive in hospital formulas, infant formulas, and feed products as a means to reduce swelling resulting from surgery or illness. Research has shown that feeding human and cow colostrum containing lysozyme to infants and calves can reduce the incidence of gastro-intestinal infections. N-acetyl-ß-D-glucosaminadase activity assays are already being used in commercial mastitis tests as an indicator for tissue damage.

REFERENCES

Adamski, F. M., King, A. T., and Demmer, J. 2000. Expression of the Fc receptor in the mammary gland during lactation in the marsupial Trichosurus vulpecula (brushtail possum). *Mol. Immunol.* 37:435–444.

Akira, S., and Takeda, K. 2004. Toll-like receptor signalling. *Nat. Rev. Immunol.* 4:499–511.

Apodaca, G., Katz, L. A., and Mostov, K. E. 1994. Receptor mediated trancytosis of IgA in MDCK cells is via apical recycling endosomes. *J. Cell Biol.* 125:67–86.

Auldist, M. J., Walsh, B. J., and Thomson, N. A. 1998. Seasonal and lactational influences on bovine milk composition in New Zealand. *J. Dairy Res.* 65:401–411.

Barrington, G. M., McFadden, T. B., Huyler, M. T., and Besser, T. E. 2001. Regulation of colostrogenesis in cattle. *Livest. Prod. Sci.* 70:95–104.

Barrington, G. M., Besser, T. E., Davis, W. C., Gay, C. C., Reeves, J. J., and McFadden, T. B. 1997. Expression of immunoglobulin G1 receptors by bovine mammary epithelial cells and mammary leukocytes. *J. Dairy Sci.* 80:86–93.

Boots, J.-W., and Floris, R. 2006. Lactoperoxidase: From catalytic mechanism to practical applications. *Int. Dairy J.* 16:1272–1276.

Bush, L. J., and Staley, T. E., 1980. Absorption of colostral immunoglobulins in newborn calves. *J. Dairy Sci.* 63:672–680.

Butler, J. E. 1973. Synthesis and distribution of immunoglobulins. *J. Am. Vet. Med. Assoc.* 163:795–798.

Chandan R. C. (2000). Functional foods and bioactive dairy ingredients. *Proc. XXX Dairy Industry Conference.* Calcutta, Dec. 8-9, pp 108-119.

Cheng, J. B., Wang, J. Q., Bu, D. P., Liu, G. L., Zhang, C. G., Wei, H. Y., Zhou, L. Y., and Wang, J. Z. 2008. Factors affecting the lactoferrin concentration in bovine milk. *J. Dairy Sci.* 91:970–976.

Christofferson, R. E. 2006. Antibiotic–An investment worth making? *Nature Biotech.* 24:1512–1514.

Clark, D. A., Phyn, C. V., Tong, M. J., Collis, S. J., and Dalley, D. E. 2006. A systems comparison of once- versus twice-daily milking of pastured dairy cows. *J. Dairy Sci.* 89:1854–1862.

Dubois, M., Gills, K. A., Hamilton, J. K., Robert, F. A. and Smith, F. (1956) *Anal. Chem.,* 28, 350.

Farr, V. C., Prosser, C. G., Clark, D. A., Tong, M., Cooper, C. V., Willix-Payne, D., and Davis, S. R. 2002. Lactoferrin concentration is increased in milk from cows milked once daily. *Proc. N. Z. Soc. Anim. Prod.* 62:225–226.

Fonteh, F. A., Grandison, A. S., and Lewis, M. J. 2002. Variations of lactoperoxidase activity and thiocyanate content in cows' and goats' milk throughout lactation. *J. Dairy Res.* 69:401–409.

Gopal, P. K., and Gill, H. S. 2000. Oligosaccharides and glycoconjugates in bovine colostrum and milk. *Br. J. Nutr.* 84:S69–S74.

Griesbeck-Zilch, B., H. Meyer, H. D., Kühn, C., Schwerin, M., and Wellnitz, O. 2008. Staphylococcus aureus and Escherichia coli cause deviating expression profiles of cytokines and lactoferrin messenger ribonucleic acid in mammary epithelial cells. *J. Dairy Sci.* 91:2215–2224.

Hancock, J. T., Salisbury, V., Ovejero-Boglione, M. C., Cherry, R., Hoare, C., Eisenthal, R., and Harrison, R. 2002. Antimicrobial properties of milk: Dependence on presence of xanthine oxidase and nitrite. *Antimicrob. Agents Chemother.* 46:3308–3310.

Hodgkinson, A. J., and Hodgkinson, S. C. 2003. Processes for production of immunoglobulin A in milk. *Agresearch Limited, assignee.* US Pat. No. 6,616,927.

Hodgkinson, A. J., Cannon, R. D., Holmes, A. R., Fisher, F. J., and Willix-Payne, D. J. 2007. Production from dairy cows of semi-commercial quantities of milk-protein concentrate (MPC) containing efficacious anti-Candida albicans IgA antibodies. *J. Dairy Res.* 74:269–275.

Hooper, L. V., Stappenbeck, T. S., Hong, C. V., and Gordon, J. I. 2003. Angiogenins: A new class of microbicidal proteins involved in innate immunity. *Nat. Immunol.* 4:269–273.

Horton, B. S. 1995. Commercial utilization of minor milk components in the health and food industries. *J. Dairy Sci.* 78:2584–2589.

Komine, Y., Komine, K., Kenzou, K., Itagaki, M., Kuroishi, T., Aso, H., Obara, Y., and Kumagai, K. 2006. Effect of combination therapy with lactoferring and antibiotics against staphylococcal mastitis on drying cows. *J. Vet. Med. Sci.* 68:205–211.

Korhonen, H. and Pihlanto, A. 2006. Bioactive peptides: Production and functionality. *Int. Dairy J.,* 16: 945-960.

Lacy-Hulbert, S. J., Woolford, M. W., Nicholas, G., Prosser, C. G., and Stelwagen, K. 1999. Effect of milking frequency and pasture intake on milk yield and composition of late lactation cows. *J. Dairy Sci.* 82:1232–1239.

Larson, B. L., Heary Jr., H. L., and Devery, J. E. 1980. Immunoglobulin production and transport by the mammary gland. *J. Dairy Sci.* 63:665–671.

Leitner, G., Krifucks, O., Jacoby, S., Lavi, Y., and Silanikove, N. 2008. Concentrations of ganglioside type M1 and immunoglobulin G in colostrum are inversely related to bacterial infection at early lactation in cows. *J. Dairy Sci.* 91:3337–3342.

Mayer, B., Doleschall, M., Bender, B., Bartyik, J., Bosze, Z., Frenyo, L. V., and Kacskovics, I. 2005. Expression of the neonatal Fc receptor (FcRn) in the bovine mammary gland. *J. Dairy Res.* 72(Special Issue):107–112.

Miles, M. 2004. *Investigation of lopopolysaccharide-binding protein in the bovine mammary gland.* MS Thesis. The University of Waikato, Hamilton, New Zealand.

Mishra V., and Prasad D. N. (2000). Probiotics and their potential health benefits. *Indian Dairyman.* 52(4): 7-13.

Mitra, A. K., Mahalanabis, D., Ashraf, H., Unicomb, L., Eeckles, R., and Tzipori, S. 1995. Hyperimmune cow colostrum reduces diarrhoea due to rotavirus: A double-blind, controlled clinical trial. *Acta Paediatr.* 84:996–1001.

Mookherjee, N., and Hancock, R. E. W. 2007. Cationic host defence peptides: Innate immune regulatory peptides as a novel approach for treating infections. *Cell. Mol. Life Sci.* 64:922–933.

Moore, M., J. Tyler, W., Chigerwe, M., Dawes, M. E., and Middleton, J. R. 2005. Effect of delayed colostrum collection on colostral IgG concentration in dairy cows. *J. Am. Vet. Med. Assoc.* 226:1375–1377.

Nagasawa, T., Kiyosawa, I., and Kuwahara, K. 1972. Amounts of lactoferrin in human colostrum and milk. *J. Dairy Sci.* 55:1651.

Nattress, F. M., and Baker, L. P. 2003. Effects of treatment with lysozyme and nisin on the microflora and sensory properties of commercial pork. *Int. J. Food Microbiol.* 85:259–267.

Newburg, D. S. 2005. Innate immunity and human milk. *J. Nutr.* 135:1308–1312.

Ormrod, D. J., and Miller, T. E. 1992. A low molecular weight component derived from the milk of hyperimmunised cows suppresses inflammation by inhibiting neutrophil emigration. *Agents and Actions* 37:70–79.

Oviedo-Boyso, J., aldes-Alarcón, J. J., Cajero-Juárez, V. M., Ochoa-Zarzosa, A., López-Meza, J. E., Bravo-Patiño, A., and Baizabal-Aguirre, V. M. 2007. Innate immune response of bovine mammary gland to pathogenic bacteria responsible for mastitis. *J. Infect.* 54:399–409.

Palmer, D. J., Kelly, V. C., Smit, A.-M., Kuy, S., Knight, C. G., and Cooper, G. J. 2006. Human colostrum: Identification of minor proteins in the aqueous phase by proteomics. *Proteomics* 6:2208–2216.

Petitclerc, D., Lauzon, K., Cochu, A., Ster, C., Diarra, M. S., and Lacasse, P. 2007. Efficacy of a lactoferrin-penicillin combination to treat β-lactam-resistant Staphylococcus aureus mastitis. *J. Dairy Sci.* 90:2778–2787.

Phalipon, A., Cardona, A., Kraehenbuhl, J. P., Edelman, L., Sansonetti, P. J., and Cotlhesy, B. 2002. Secretory component: A new role in secretory IgA-mediatedimmune explusion in vivo. *Immunity* 17:107–115.

Rainard, P., and Riollet, C. 2006. Innate immunity of the bovine mammary gland. *Vet. Res.* 37:369–400.

Rincheval-Arnold, A., Belair, L., and Djiane, J. 2002. Developmental expression of pIgR gene in sheep mammary gland and hormonal regulation. *J. Dairy Res.* 69:13–26.

Sanchez, L., Lujan, L., Oria, R., Castillo, H., Perez, D., Ena, J. M., and Calvo, M.1992. Synthesis of lactoferrin and transport of transferrin in the lactating mammary gland of sheep. *J. Dairy Sci.* 75:1257–1262.

Schlimme, E., Meisel, H. 1995. Bioactive peptides derived from milk proteins. Structural, physiological and analytical aspects. *Nahrung.* 1995;39(1):1-20.

Sordillo, L. M., Shafer-Weaver, K., and DeRosa, D. 1997. Immunobiology of the mammary gland. *J. Dairy Sci.* 80:1851–1865.

Vermeirssen V, Van Camp, J, Decroos, K, Van Wijmelbeke, L., and Verstraete, W. (2003a). *The impact of fermentation and in vitro digestion on the formation of ACE inhibitory activity.*

Willix-Payne, D., Prosser, C. G., Hodgkinson, A. J., Hutchinson, J., Cannon, R. I., and Fisher, F. J. 2001. Novel Immunoglobulins. In *Proc. Global Bioactives Summit,* Hamilton, New Zealand. CD-ROM.

In: Recent Advances in Microbiology. Volume 3 ISBN: 978-1-53614-057-6
Editors: N.C. Gautam, S.P. Tiwari and R. Sharma

Chapter 6

HOW PSYCHROPHILES SURVIVE COLD: MOLECULAR MECHANISMS AND INDUSTRIAL APPLICATIONS

Prachi Bhargava[1], Ravindra Soni[1] and Reeta Goel[2,*]
[1]Amity Institute of Biotechnology, AUUP, Noida, Uttar Pradesh, India
[2]Department of Microbiology, CBSH, G. B. Pant University of Agriculture and Technology, Pantnagar, Uttaranchal, India

INTRODUCTION

Microbial life has evolved the capacity to thrive in a broad range of thermal environments. Thermal adaptation, in particular to extreme conditions, has restricted the range of tolerable temperatures (Siddiqui & Cavicchioli, 2006). It seems to be a fundamental law that, wherever microbial life can survive, it will be found to exist as they are the most hardiest, spore forming organisms which can grow and reproduce at temperatures less than 0°C in high-altitude cloud droplets. The largest proportion of biomass on earth is generated at cold temperatures (≤5°C). These environments comprise, among others, the Antarctic, Arctic and mountain regions, as well as the deep-sea waters that cover three-quarters of the planet surface. This is mainly due to the contribution of vast numbers of microorganisms in the oceans, although the cold biosphere extends to permanently cold alpine regions, caves, upper atmosphere, and polar regions, in addition to seasonally cold environments (Luisa et al. 2009).

[*]Corresponding Author Email: rg55@rediffmail.com.

Evolution has allowed these cold adapted organisms, called psychrophiles, not to merely survive, but to breed and grow successfully in the restrictive conditions of cold habitats.

Far less is known about psychrophilic bacteria than about thermophiles, and it is not even certain yet whether life on earth originated in a hot or cold environment.

Psychrophiles display metabolic fluxes at low temperatures that are more or less comparable to those exhibited by closely related mesophiles living at moderate temperatures (Deming, 2002; Russell, 2000), clearly showing that mechanisms of temperature adaptations are involved.

Cold-adapted microorganisms can be categorized as psychrophilic (an organism with an optimal growth temperature at or below 15°C, and a maximum growth temperature below 20°C), psychrotrophic (an organism exhibiting the ability to grow at temperatures below 15°C but exhibiting maximum growth rates at temperature optima above 18°C) organisms, and are often subjected to other extreme environmental parameters and psychrotolerant (organisms tolerant of low growth temperatures). Some are specialized to very low temperatures (steno-psychrophilic bacteria, with optimal growth temperature <15°C), and some tolerate temperature up to 20°C (eury-psychrophiles). (Siddiqui et al., 2006). The differentiation between psychrotrophs and psychrophiles is, by and large, ignored now-a-days and all the cold-tolerant bacteria are called psychrophiles (Chattopadhyay, 2006).

These microorganisms are devoid of any temperature regulation, their internal temperature synchronizes to that of the surrounding environment. They have very strongly challenged the strong negative effect of low temperatures on biochemical reactions, and continue to breed, grow and move at rates similar to those achieved by closely related species living in temperate environments (Charles, 2000).

Lot of important and essential adaptations are required to survive at such low temperatures, including the expression of enzymatic activities at appropriate levels as well as the maintenance of membrane stability and permeability, the latter being essential for many of the major cellular functions such as nutrient uptake, passive and active transport of various compounds, electron and proton transport, photosynthesis, environmental sensing, and recognition processes. The changes at the level of three dimensional structure of protein are incorporated as mutations or modifications at the gene level. These differences via transcription and translation form an altogether differently modified pattern of proteins expressed through the thermal regulation of gene expression (Zecchinon, 2001).

COLD ADAPTED ENZYMES

Enzymes, which are proteins capable of catalyzing all the biochemical reactions occurring within an organism that render them compatible with life, are an essential target for the adaptation of an organism to a cold environment (Gerday, 1997).

Psychrophilic microorganisms, hosts of permanently cold habitats, produce enzymes which are adapted to work at low temperatures and display a higher catalytic efficiency over a temperature range of roughly 0-30ºC and a high thermos sensitivity when compared to their mesophilic counterparts (Gerday, 1997). Owing to their useful and unique catalytic properties in industrial applications, a great effort has been put in discovering and studying more novel psychrophilic enzymes. For example, lipases and esterases potentially catalyze region and stereo-selective hydrolysis reactions and can exhibit organic solvent-stable catalytic activities (Jensen & Hamosh, 1969; Klibanov, 2000; Lang & Dijkstra, 1998).Recently, there have been bans on certain animal derived products example pig liver esterase which were been used extensively in pharma industry, there was a sudden growth in the use of microbial enzymes. There a vast gap between the potential of microbial enzymes to be used and exploring them. Using metagenomics, many novel biocatalysts have been screened for and identified, such as chitinase (LeCleir, 2007), protease (Waschkowitz et al. 2009), oxidase (Ye et al. 2010), oxygenase (Singh et al. 2006), amylase (Sharma et al. 2010), xylanase (Hu et al. 2008), lipase/esterase (Rhee et al. 2005), lysine racemase (Chen et al. 2009), and polyketide synthase (Chung et al. 2008).

CHALLENGES IN COLD REGIONS

1) Most of the physiological processes and membrane fluidity are reduced which provokes an increased viscosity of water.
2) A decrease in temperature also induces a reduction in salt solubility and an increase in gas solubility.
3) With the decreasing temperatures, the pH of the buffers also decreases, affecting both the protein solubility and the charge of amino acids, particularly histidine residues.
4) Enzymes are also subject to cold denaturation, leading to the loss of enzyme activity at low temperatures (Privalov, 1990). This phenomenon is thought to occur through the hydration of polar and non-polar groups of proteins (Makhatadze & Privalov, 1995) a process thermodynamically favoured at low temperatures. It is notably the weakening of hydrophobic forces, which are crucial for protein folding and stability, which causes protein unfolding.

5) Another consequence of the exposure to low temperatures is a strong inhibition of chemical reaction rates catalysed by enzymes. The temperature dependence of chemical reactions is commonly described by the Arrhenius equation k ¼ Ae_Ea=RT, in which k is the rate constant, A is the pre-exponential factor (related to steric factors and molecular collision frequency), Ea is the activation energy, R is the gas constant (8.314 kJ mol_1) and T is the absolute temperature in Kelvin. Thus, any decrease in temperature will induce an exponential decrease in the reaction rate.
6) Other major changes include reduced activity of transcriptional and translational enzymes; reduced protein folding, owing primarily to a reduced rate of prolyl isomerization; and a stabilization of DNA and RNA secondary structures.

Facing the Challenges for Cold Adaptation

The fight against low temperatures have made psychrophiles to adopt various strategies beginning right at the level of individual molecules to that of the whole organism. They have to overcome the detrimental challenges like reduced enzyme activity; decreased membrane fluidity; altered transport of nutrients and waste products; decreased rates of transcription, translation and cell division.

These organisms have several mechanisms to face the above adversities like the synthesis of specialized molecules known as cold-shock proteins, cryoprotectors and antifreeze molecules to fight against protein cold-denaturation; inappropriate protein folding; and intracellular ice formation. They have regularized the permeability of ion channels, polymerized microtubules, undergo seasonal dormancy, and modified enzyme kinetics (Georlette et al. 2004).

Important cellular functions, such as passive and active permeability, nutrient uptake, electron transport, environmental sensing and recognition processes require the maintenance of membrane stability. Psychrophiles achieve this, notably by increasing the proportion of unsaturated fatty acids and the modulation of the activity of the enzymes involved in fatty acid and lipid biosynthesis (Margesin et al. 2002). The effect of temperature on chemical reactions is basically described by the Arrhenius equation: *k* = *Ae_Ea/RT*, where *k* is the rate constant, *A* is the pre-exponential factor that depends on the reaction, *E*a is the so-called activation energy, *R* is the gas constant (8.31 kJ mol_1) and *T* is the temperature in kelvins. Accordingly, any decrease in temperature will induce an exponential decrease of the reaction rate, the extent of which depends on the value of the activation energy. The thermodependence of the activity can be approximately expressed by the *Q*10 value that is normally close to 2– 3. This is the main factor preventing the growth, at low temperatures, of non-adapted organisms (Salvino et al. 2002).

The best possible way to maintain sustainable activity at a permanently low temperature is to produce a cold adapted enzyme with enhanced catalytic efficiency *k*cat/*K*m as producing more enzyme to counteract the negative effect of cold on the activity of an enzyme would be energetically expensive. Generally, cold-adapted enzymes optimize their catalytic efficiency by increasing *k*cat, decreasing *K*m or by changes in both parameters. The rate of all reactions including enzymatic reactions is described by the Arrhenius equation

$$kcat = \mathrm{A}\kappa e - Ea/\mathrm{R}T$$

where *kcat* is the enzyme reaction rate, which increases with an increase in absolute temperature (T) and a decrease in activation energy (Ea), A is the preexponential factor, κ is the dynamic transmission coefficient (generally assumed to be 1), and R is the universal gas constant (8.314 J mol−1 K−1) (Feller & Gerday, 1997, Smalas et al. 2000). Some psychrophiles express some isoenzymes (Baldwin & Hochachka, 1970; Jagdale & Gordon, 1997; Somero, 1995, or follow the strategy of energetically expensive strategy of increasing enzyme concentration (Crawford & Powers, 1992; Devos, 1998) to compensate for the very slow metabolic rates that would occur at low temperatures as a result of this kinetic effect.

The comparison of the crystal structures of cold adapted enzymes and their homologous enzymes from mesophilic and/or thermophilic organisms is yielding fruits to elucidate the structural basis of temperature adaptation. Structural comparisons support the original hypothesis that cold-adapted proteins appear to contain fewer or weaker intra molecular interactions, which is assumed to facilitate global or local flexibilities in the protein molecules. The most frequently reported structural differences involve interactions such as fewer intra- or inter-subunit salt bridges, less compact hydrophobic packing in the protein core, longer surface loops and fewer prolines in such loops, an increased number of glycine clusters, a reduced number of proline/arginines, improved solvent interactions through additional surface (especially negative) charges, increased solvent exposure of apolar surfaces, and a better accessibility of the active site.

STRUCTURAL ARCHITECTURE

The unique properties of cold-adapted enzymes, i.e., a high specific activity at low temperatures and a high thermosensitivity, can be useful in various biotechnological processes (Demirjian, 1998; Gerday et al. 2000; Russell, 1998). The first crystallographic 3D structure of a cold enzyme isolated from a microorganism, an a-amylase isolated from Gram negative Antarctic bacteria, was described in 1996 (Aghajari et al. 1996).

Crystallographic structures of a number of cold- adapted enzymes have revealed that only minor structural modifications are needed to adapt a mesophilic homologue to cold temperatures and all active-site residues involved in the reaction mechanisms are strictly conserved between homologous enzymes adapted to different temperatures. It was found that all structural factors currently known to stabilize the mesophilic protein counterpart could be attenuated in strength and number in cold-active enzymes. This could possibly involve an increased number and clustering of glycine residues of forming multiple (providing local mobility); a decrease in proline residues in loops (providing enhanced chain flexibility between secondary structures); a reduction in arginine residues, capable electrostatic interactions and hydrogen bonds, and eventually a lowering of the number of ion pairs, aromatic interactions, an increased exposure of hydrophobic residues to solvent (entropy-driven destabilizing factor) (Feller & Geraday, 1997; Feller et al. 1997). Obviously, each protein uses a few of these structural alterations in order to acquire the required flexibility to be more or less adapted to the temperature of the environment. Interestingly, the same structural factors have been implicated in the stability of thermophilic proteins (Jaenicke, 1991; Jaenicke et al. 1996; Vieille & Zeikus, 2001) suggesting that there is a continuum in the strategy of protein adaptation to temperature.

Activity–Stability–Flexibility Relationships

Thermal stability of enzymes is inversely proportional to the structural rigidity, thus thermophilic enzymes are poor catalysts at room temperature since the structural rigidity restricts the interaction between enzyme and its substrate. On the contrary, cold adaptability requires structural flexibility, which favours greater complementarity at low energy cost thus explaining the high specific activity of some cold-adapted enzymes (Gerday et al. 2000). Low temperatures tend to improve the compactness of a protein by limiting the 'breathing' of the structure corresponding to micro-unfolding processes. Therefore, at low temperatures, a mesophilic protein will lose the mobility required for its catalytic activity. The current accepted hypothesis (Gerady et al. 1997; Zavodszky, 1998) suggests that psychrophilic enzymes have to decrease their structural rigidity in order to perform catalysis at low temperatures, the enhanced plasticity being generated by the generally low stability of the protein structure. An optimum balance between flexibility and stability is the key point in the adaptation of a protein to environmental temperature. A significant increase in the overall flexibility of the protein could possibly lead to enzymes displaying a high specific activity at low temperature but also a weak affinity for the substrates and a very low stability.

The type and strength of interactions between hydrophobic residues (core hydrophobicity) and between hydrophobic residues and solvent water molecules (surface hydrophobicity) are the governing factors which decide the structural flexibility and thermolability of a cold adapted protein The buried amino acids in cold-adapted enzymes tend to be smaller and less hydrophobic than in homologs from mesophiles and thermophiles. Fewer Ile residues are present in the inner core of cold-adapted trypsins since due to its branching and size, Ile can pack more efficiently inside the core and stabilize a protein (Agrahari et al. 1996). The surfaces of cold-adapted enzymes tend to have a higher proportion of hydrophobic (nonpolar) residues (Agrahari et al. 1996; Gianese et al. 2002).

Hydrophobic surface residues will destabilize a protein structure because of the decreased entropy of water molecules, which form cage-like structures around nonpolar residues. However, at low temperatures, the entropy gain is reduced owing to the decreased mobility of the released water molecules (Kumar et al. 2004; Kumar et al. 2002). This implies that cold-adapted enzymes may gain flexibility from, and have a greater capacity to tolerate, increased surface hydrophobicity (Saunders et al. 2003). Cold adaptation appears to involve a strategic arrangement of the total exposed and buried nonpolar fraction, leading to increased exposure of the nonpolar residues to water, with a concomittant reduction in the packing of the hydrophobic core (Gianese et al. 2002).

A reduced Arg/Lys ratio peculiarly signifies a cold adapted enzyme as arginine residues enhance enzyme thermostability more than Lys residues by facilitating a greater number of electrostatic (up to two salt bridges and five H-bond) interactions through their guanidino group. The polarity of aromatic amino acids also permits favorable interactions between aromatic rings at right angles to each other (aromatic-aromatic interactions) or between aromatic rings and the side chains of Arg and Lys (aromatic-amino interactions). Aromatic interactions may therefore promote thermostabilization through an enthalpic contribution. Reduced number of aromatic interactions appear to be important for maintaining flexibility. Longer surface loops increase the possible amplitude of the movement between secondary structures and may decrease enzyme stability.

With the decrease in environmental temperature, a number of changes are known to occur in the fatty acid profile of bacterial cell membrane to maintain an optimum fluidity.

Conversion of saturated fatty acids into unsaturated fatty acids by the enzymes desaturases, induced at low temperature and preferential synthesis of short-chain fatty acids, branched chain fatty acids and anteiso-fatty acids to long chain fatty acids, straight-chain fatty acids and iso fatty acids respectively are well-documented in the literature. Conversion of *cis* fatty acids into *trans* fatty acids, by the enzyme *cis-trans* isomerase, is believed to facilitate the survival of bacteria at higher temperature, since the *trans* fatty acids are known to contribute to the decrease in membrane fluidity.

Studies suggest that in response to the increase in the synthesis of membrane-fluidizing fatty acids, synthesis of membrane-rigidifying polar carotenoids was also enhanced to counterbalance the effects of fatty acids in the Antarctic bacteria (Chattopadhyay, 2006).

COLD SHOCK PROTEINS

Sudden decreases in temperature will also elicit a specific alteration in gene expression known as the coldshock response (Cavicchioli et al. 2000; Ermolenko & Makhatadze, 2002; Graumann & Marahiel, 1996; Jones & Inouye, 1969; Phadtare & Alsina, 1999). Notably, all organisms from prokaryotes to plants and higher eukaryotes respond to cold shock in a comparatively similar manner. The cold-shock response, which is evidently not confined to psychrophilic organisms but constitutes the beginning of cold-adaptation, induces the synthesis of cold-shock proteins named Csp. Cold shock proteins (Csps) are a family of small sized proteins that have a highly conserved structure and bind to single-stranded nucleic acids via their nucleic acid binding motifs RNP1 and RNP2. These proteins act mainly on the regulation of cellular protein synthesis, particularly at the level of transcription and the initiation of translation; and they also act as chaperone by preventing the formation of mRNA secondary structures (mRNA _folding). The regulation of cold-shock-protein itself is a complex and multifactorial phenomenon, being controlled at the level of both transcription and translation. A major distinction between psychrophiles and their mesophilic and thermophilic homologues is that, in the former, the synthesis of housekeeping gene products is not inhibited by cold-shock, and the number of coldshock proteins is usually higher, and proportional to the severity of the cold-shock (Berger et al. 1996; Mayr et al. 1996).

Bacterial Csps are mainly induced after a rapid temperature downshift to regulate the adaptation to cold stress, but are also present under normal conditions to regulate other biological functions. The structural unit characteristic for Csps occurs also as a cold shock domain (CSD) in other proteins and can be found in wide variety of organisms from bacteria to vertebrates. Important examples are the Y-box proteins that are known to be involved in regulation of several transcription and translation processes.

A number of plant genes are induced by low-temperature stress, and, in prokaryotes, cold stress induces several well-characterized CSPs (Fujita, 1999). In contrast, there is sparse information on the molecular mechanisms of the cold-shock response in mammalian cells; however, our current understanding suggests that the response involves a coordinated series of events involving modulation of transcription, translation, the cell cytoskeleton, the cell cycle and metabolic processes. The prokaryotic and eukaryotic cells generally suppress transcription and subsequent translation, in response to cold stress, except for a select number of CSPs whose synthesis continues or is up-regulated during cold shock (Ermolenko & Makhatadze, 2002). In *Escherichia coli*, approx. 27 CSPs have been identified out of which the majority of CSPs are regulated sequentially in response to temperature downshift and have fundamental functions that determine cell fate such as DNA replication, transcription, translation, RNA stabilization and ribosome assembly (Gualerzi et al. 2003). Cold shock proteins (Csps) are a family of small sized proteins that have a highly conserved structure and bind to single-stranded nucleic acids via their nucleic acid binding motifs RNP1 and RNP2. Bacterial Csps are mainly induced after a rapid temperature downshift to regulate the adaptation to cold stress, but are also present under normal conditions to regulate other biological functions. The structural unit characteristic for Csps occurs also as a cold shock domain (CSD) in other proteins and can be found in wide variety of organisms from bacteria to vertebrates. Important examples are the Y-box proteins that are known to be involved in regulation of several transcription and translation processes.

ANTIFREEZE PROTEINS

Antifreeze proteins (AFPs) are found in some organisms such as fish, insects, plants, fungi and microorganisms that encounter freezing conditions (Jia, and Davies, 2002). The role of antifreeze proteins (AFPs) in preventing freezing of blood of fishes in low-temperature environment is well-known. By contributing to both freeze resistance and freeze tolerance, AFPs have helped to increase species diversity in some of the harshest and most inhospitable environments. For instance, fish from the polar regions survive at temperatures close to the freezing point of sea water 1.8°C by secreting high concentrations of AFPs molecules into their blood (DeVries, 1988; Cheng, 1998), preventing the formation of ice crystals which could lead to the destruction of cell membranes and the disruption of osmotic balance. Such molecules act in the blood of these fish by adsorption to the surfaces of microcrystals of ice, thereby preventing the association of additional water molecules and thus, growth of the crystals (Ewart et al. 1999; Harding et al. 1999; Verdier et al. 1996).Recent works suggest that the key feature of all antifreeze molecules could lie in the complementarity between the ice-binding protein surfaces and the ice crystals (Leinala et al. 2002). In freeze-avoiding organisms (e.g., primarily fishes) the AFPs act by promoting supercooling of the body fluids at subzero temperatures. This is called thermal hysteresis (TH). On the other hand, in freeze-tolerant organisms (e.g., some plants and insects), the AFPs inhibit the re-crystallization of the extracellular ice crystals. The role of AFPs in bacterial cold adaptation has started coming to the limelight during the past few years.

Most of the bacterial AFPs, known so far, are characterized by substantially lower TH value compared to that of AFPs, isolated from animals. Hence the strategies, adopted by the bacteria, which produce them for adaptation to low temperature, are freeze-tolerance and not freeze avoidance.

Transcription and Translation

At low temperature protein synthesis is hindered because of: reduced activity of transcriptional and translational enzymes; reduced protein folding, owing primarily to a reduced rate of prolyl isomerization; and a stabilization of DNA and RNA secondary structures. The psychrophilic enzymes are optimally active at low temperatures; a ribosomal extract, RNA polymerase, elongation factor and peptidyl–prolyl *cis–trans* isomerase have all been shown to retain activity near 0°C in several psychrophilic microorganisms. Some specific proteins like nucleic-acid-binding proteins—*Escherichia coli* CspA-related proteins—and RNA helicases that are important for the destabilization of DNA and RNA secondary structures are also overexpressed at low temperatures in psychrophiles (Berger et al. 1996).

Industrially Important Cold Active Enzymes

During recent years the biotechnological potential of cold adapted microorganisms have been recognized offering numerous economic and ecological advantages over the use of mesophiles and their enzymes which operate at higher temperatures. The cold adapted microorganisms are source of commercially and industrially important enzymes like lipase, cellulase, protease etc.

Cold Active Lipase

Lipases are hydrolases (E.C.3) which act on ester bonds (E.C.3.1) of carboxylic esters (E.C.3.1.1). They hydrolyze triacylglycerols to fatty acids, diacylglycerol, monoacylglycerol, and glycerol and are known as triacylglycerol acyl hydrolases (E.C.3.1.1.3). Lipases possess broad substrate specificity and show optimum activities over a wide range of temperature. These interesting properties make lipases the most versatile biocatalyst (Kademi et al. 2005). Cold-adapted lipolytic microorganisms produce lipases, which function effectively at cold temperatures with high rates of catalysis in comparison to the lipases from mesophiles or thermopiles, which show little or no activity at low temperature. These lipases have evolved a range of structural features that confer a high level of flexibility, low activation enthalpy, low-substrate affinity, and high specific activity at low temperatures. Moreover, the maximum level of activity of these lipases is shifted towards lower temperatures with a concomitant decrease in thermal stability.

Economic Importance of Cold Active Lipases

Lipase mediated modifications occupy a prominent place in oil industry for tailoring structured lipids since enzymatic modifications are specific and can be carried out at moderate reaction conditions (Gupta et al. 1999). Cold active esterases and lipases play a significant role of additives in detergents, mainly in household dishwashers and household and industrial laundry, clearing of drains clogged by lipids in food processing or domestic/industrial effluent treatment plants (Bailey & Ollis, 1986). Cold adapted enzymes offer a high potential in this type of application in cold environmental conditions, but production costs still limit their current use. Enzymes can reduce the environmental load of detergent products, since they save energy by enabling a lower wash temperature to be used; allow the content of other, often less desirable, chemicals in detergents to be reduced; are biodegradable, leaving no harmful residues; have no negative impact on sewage treatment processes; and do not have a risk to aquatic life. Commercial preparations used for the desizing of denim and other cotton fabrics, contains both α amylase and lipase enzymes.

Further, it is used in liquid leather cleaner (Klibanov, 2000), a bleaching composition (Nakamura & Nasu), decomposition of lipid contaminants in dry-cleaning solvents (Abo, 1990), contact lens cleaning (Bhatia, 1990), degradation of organic wastes on the surface of exhaust pipes, toilet bowls, etc. (Moriguchi et al. 1990). Removal of dirt/cattle manure from domestic animals by lipases and cellulases (Abo, 1990), washing, degreasing and water reconditioning by using lipases along with oxidoreductases, which allows for smaller amounts of surfactants and operation at low temperatures.

The market of recombinant psychrophilic lipases covers about 40% of the total sale of enzymes. In the food industry, cold-active lipases might be useful for the development of flavours owing to their high specific activity and unique features. Animal feed can also constitute a vast application field for these enzymes in terms of improvement of digestibility and assimilation. One can also apply these enzymes for the synthesis of valuable peptides, fatty acids and polysaccharides by reverse hydrolysis in low water amounts, taking advantage of the high flexibility of these enzymes to further decrease water activity and improve yields. This process can also be extended to pharmaceutical and chemical industries for the production of compounds with high added value (Luisa et al. 1990). Studies have been done to utilise cold-active lipolytic enzymes for degradation of soils and waters contaminated by hydrocarbons, heavy metals, and other xenobiotic Compounds (Margesin, 2007).The use of enzymes, especially lipases, in organic solvents proves an excellent methodology for the preparation of single-isomer chiral drugs (Gotor, 2006). Use of lipase B from *C. antarctica* for the preparation of optically active alcohols has been reported (Rotticci et al. 2001).

Cold adapted lipase is used in organic synthesis, chiral synthesis or resolution, modification of fats and oils, synthesis of sugar-based polymers, etc. The use of lipases in esterification reactions to produce industrially important products such as emulsifiers, surfactants, wax esters, chiral molecules, biopolymers, modified fats and oils, structured lipids, and flavor esters is well documented (Krishna et al. 2002). Cold-adapted lipase from Pseudomonas strain P38 is used for the synthesis in n-heptane of the flavoring compound, butyl caprylate (Tan et al. 1995).

Modifications of Cold Adapted Lipases

The lipase consists of a single domain molecule and all lipases conform to a common structural organization, viz., the alpha/beta hydrolase fold (Nardini & Dijkstra, 1999; Ollis et al. 1992). The active site of lipase contains the catalytic triad, Ser-His-Asp, common to all serine hydrolases where Ser acts as the nucleophile, His as the basic residue, and Asp or Glu as the acidic residue (Uppenberg et al. 1994b). Such a catalytic triad exists in enzymes with different folding, including trypsin and subtilisin, and is an example of convergent evolution (Holmquist, 2000). Access to the active site, consisting a serine, histidine, carboxylic acid triad may be shielded by a mobile lid, whose position closed or open determines the enzyme in an inactive or active conformation. The activation can be explained by the opening of a lid (flap) structure of the enzyme at an interface. The lipase with open lid is the active form of the enzyme and gives the substrate access to the active site. The substrate-binding site is located inside a pocket on top of the central β-sheet that is typical of this fold. Size and geometry of the substrate-binding cleft have been related to substrate specificity (Pleiss et al. 1998) and residues that contact the substrate have been identified by crystallography and docking. However, it is recognized that other protein regions, such as the lid itself (Brocca et al. 2003) and the reaction conditions (Verger, 1980) may play a role in lipases specificity.

Cold active lipases are structurally modified by an increasing flexibility of the polypeptide chain enabling an easier accommodation of substrates at low temperature. The fundamental issues concerning molecular basis of cold activity and the interplay between flexibility and catalytic efficiency are important in the study of structure–function relationships in enzymes. Such issues are often approached through comparison with the mesophilic or thermophilic counterparts, by site-directed mutagenesis and 3D crystal structures (Narinx et al. 1997; Wintrode et al. 2000). A very low proportion of arginine residues as compared to lysines, a low content in proline residues, a small hydrophobic core, a very small number of salt bridges and of aromatic–aromatic interactions are the possible features of lipase for cold adaptation. Similarly the weakening of hydrophobic clusters, the dramatic decrease (40%) of the Proline content and of the ratio Arg/Arg+Lys makes lipases active at low temperature (Gerday et al. 1997).

In addition to this, the structural factors possibly involved in cold adaption are increased number and clustering of glycine residues (providing local mobility), lower number of ion pairs and weakening of charge-dipole interactions in α helices (Georlette et al. 2004; Gomes & Steiner, 2004). The substitution of Glycine with Proline by mutation caused a shift of the acyl chain length specificity of the enzyme towards short-chain fatty acid esters and enhanced themostability of the enzyme (Kulakova et al. 2004). Introduction of polar residues in the surface of exposed lid might be involved in improved substrate specificity and protein flexibility. These structural adaptations may confer on the enzyme a more flexible structure, in accordance with its low activation energy and its low thermal stability.

Cold Active Cellulases

Cellulases (EC 3.2.1.4) are endoenzymes that hydrolyse randomly *β*-1, 4-glycosidic bonds in cellulose, releasing cellobiose as the smallest product. These enzymes can be found associated with the cellulosome, an extracellular supramolecular assembly produced by some bacteria and fungi (Carrard et al. 2000; Shoham et al. 1999), but are generally secreted as independent enzymes by cellulolytic microorganisms. The catalytic domain of cellulases is frequently connected to a CBM (cellulose-binding module) via a linker region (Tomme et al. 1995). Cellulose, which is one of the most abundant carbohydrates, is found in nature almost exclusively in plant cell walls, although it is produced by some animals (e.g., tunicates) and a few bacteria, provides the major renewable energy source on the planet Earth. Its turnover, therefore, plays an important role in the global carbon cycle for all living organisms. Cellulose synthesis and recycling in the cold environments account for a large proportion of the carbon cycle because approximately 80% of the biosphere and more than 90% of the marine environments have temperatures lower than 5°C. Cellulose synthesized by the aerobic bacterium *Acetobacterxylinum* has been tremendously useful as a model system for studying cellulose biosynthesis, but has only been used for a few studies of microbial cellulose utilization. Like plant cellulose, bacterial cellulose is highly crystalline, but the two celluloses differ in the arrangement of glucosyl units within the unit cells of the crystallites (Atalla & Vanderhart, 1984) and genetic evidence suggests that the two celluloses are synthesized by enzymatic machinery that differs considerably at the molecular level (Brown et al. 2000). The two celluloses also differ substantially in rate of hydrolysis by fungal cellulases (Hayashi et al. 1997) and in rate of utilization by mixed ruminal bacteria (Schofield et al. 1994; Weimer et al. 2000). The variable structural complexity of pure cellulose and the difficulty of working with insoluble substrates has led to the wide use of the highly soluble cellulose ether, carboxymethylcellulose (CMC), as a substrate for studies of endoglucanase production. Unfortunately, the use of CMC as an enzymatic substrate has weakened the meaning of the term 'cellulolytic,' since many organisms that cannot degrade cellulose can hydrolyze CMC via mixed-glucan enzymes (Fields et al. 1998). Because of the substituted nature of the hydrolytic products, relatively few microbes (including some fungi and *Cellulomonas* strains) can use CMC as a growth substrate.

The genes encoding cellulases are chromosomal in both bacteria and fungi. In the fungi, cellulase genes are usually randomly distributed over the genome, with each gene having its own transcription regulatory elements. As the crystal structures of llulases are being solved, this will considerably facilitate the design of rational engineering strategies for enzymes with desired properties (enzymes with altered specificity, stability, biophysical properties).The results of small-angle X-ray scattering showed that the linker region between the catalytic binding module and the cellulose-binding module is unstructured and unusually long and flexible compared to its mesophilic counterpart, indicating that the linker plays a significant role in the cold adaptation of this psychrophilic cellulase. The longer linker in the cellulase from the psychrophilic *P. haloplanktis* as compared to the cellulose from *Erwinia chrysanthemi* is a highly flexible structure due to various factors, like negatively charged residues, the absence of positively charged residues, and only a small number of hydrophobic residues. Due to the highly flexible structure of the linker, the catalytic module and the cellulose-binding domain are free to move to a larger extent, which helps in their correct positioning with respect to the cellulose substrate at low temperature below 25°C.

An inspection of these trees reveals that the ability to digest cellulose is widely distributed among many genera in the domain *Bacteria* and in the fungal groups within the domain *Eucarya*, although no cellulolytic members of domain *Archaea* have yet been identified. Within the eubacteria there is considerable concentration of cellulolytic capabilities among the predominantly aerobic order *Actinomycetales* (phylum *Actinobacteria*) and the anaerobic order *Clostridiales* (phylum *Firmicutes*). Fungal cellulose utilization is distributed across the entire kingdom, from the primitive, protist-like Chytridomycetes to the advanced Basidiomycetes.

Members of genera that have received considerable study with respect to their cellulolytic enzymes and/or wood-degrading capability include *Bulgaria, Chaetomium*, and *Helotium* (Ascomycetes); *Coriolus, Phanerochaete, Poria, Schizophyllum* and *Serpula* (Basidiomycetes); and *Aspergillus, Cladosporium, Fusarium, Geotrichum, Myrothecium, Paecilomyces, Penicillium*, and *Trichoderma* (Deuteromycetes). Natural sources of cellulases generally contain at least three groups of cellulolytic activities, including endoglucanase, β-1,4-cellobiohydrolase, and β-glucosidase. The optimum temperature for their activity generally ranges from 20 to 40°C (Table 2), with one study reporting a higher optimum temperature for activity (50°C) by a cellulose from the psychrotrophic yeast *Rhodotorulaglutinis.*

Most of the cellulases produced by psychrophilic microorganisms showed optimum activity in the acidic to neutral pH range (pH 4.5–7.0), with the exception of cellulases from *Paenibacillus* sp., *Pseudoalteromonas*sp. MB-1, and *Shewanella* sp., which showed their optimum activity in the neutral to alkaline pH range (Kasana & Gulati, 2011). Cellulases from psychrophilic microorganisms reported up to now have a size range of 40–93 kDa.

Applications of Cellulases

Cellulases have a wide range of applications. The main potential applications are in food, animal feed, textile, fuel and chemical industries. Other areas of application include the paper and pulp industry, waste management, medical/pharmaceutical industry, protoplast production, genetic engineering and pollution treatment (Garsoux et al. 2004; Papaleo et al. 2008). In food industry, cellulases are being used (a) in the extraction of fruit juices and oil from seeds, (b) in the clarification of fruit juices, (c) to increase the soaking efficiency and homogeneous water absorption of cereals, (d) in the removal of external soybean coat in production of fermented soybean foods such as soy sauce and miso, (e) in the isolation of proteins from soybean and coconut, (f) for the efficient isolation of starch from corn and sweet potato, (g) for the gelatinization of seaweeds to increase digestibility, (h) for the extraction of agar from seaweeds, and (i) to digest ball-milled lignocellulose which can be used as food additive (Beguin & Anbert, 1993; Mandels, 1985). Cellulases can also be used for (a) improving the nutritive quality of fermented foods, (b) improving the rehydrability of dried vegetables and soup mixtures, (c) the production of cello-oligosaccharides, glucose and other soluble sugars from cellulosic wastes, and (d) for the removal of cell wall which will facilitate the release of flavors, enzymes, polysaccharides and proteins (Mandels, 1985).

Cellulases have been incorporated into detergents since the early 1990s for laundry purposes. During textile washing, cellulases remove cellulose microfibrils formed during manufacturing and washing of the cotton based cloth (Brown et al. 1986). Cellulases can also be used as color brightening and softening agents. Considering the present market scenario in the detergent industry, incorporating cellulases with proteases in detergent formulations would be advantageous for enhancing the washing performance of detergents. The detergent industry is now focusing on cellulases for lower (cooler) washing temperatures (30-40°C) and reduction in water consumption. Hence, for using the cellulases as detergent additives, the enzyme should be active under low temperature, stable under alkaline conditions, and compatible with oxidants and detergents. If the cellulose is resistant to degradation (hydrolysis) by proteases, that would be of added advantage for detergent formulation and reduce the overall cost of detergent formulation. The comparison of the crystal structures of cold adapted enzymes and their homologous enzymes from mesophilic and/or thermophilic organisms is yielding fruits to elucidate the structural basis of temperature adaptation. Structural comparisons support the original hypothesis that cold-adapted proteins appear to contain fewer or weaker intramolecular interactions, which is assumed to facilitate global or local flexibilities in the protein molecules. The most frequently reported structural differences involve interactions such as fewer intra- or inter-subunit salt bridges, less compact hydrophobic packing in the protein core, longer surface loops and fewer prolines in such loops, an increased number of glycine clusters, a reduced number of proline/arginines, improved solvent interactions through additional surface (especially negative) charges, increased solvent exposure of apolar surfaces, and a better accessibility of the active site. As the crystal structures of cellulases are being solved, this will considerably facilitate the design of rational engineering strategies for enzymes with desired properties (enzymes with altered specificity, stability, biophysical properties).

Cold Active/Adapted Proteases

Proteases from microorganisms represent one of the three largest groups of industrial enzymes accounting for 60% of the total worldwide sale of enzymes. Proteases, on the basis of the position where they cleave can be divided in exopeptidases (cleaves the peptide bond proximal to the amino or carboxy terminus of the substrate) or endopeptidases (cleaves peptide bonds distant from the termini of a substrate).

In order to have an indepth knowledge of the molecular basis of adaptation to low temperature, structure of psychrophilic proteases were studied. Evolution played a major role in the development of essential structural features of psychrophilic proteases that contribute to their high psychrophilic activity. Cold adapted proteases probably are structurally modified by an increasing flexibility of the polypeptide chain enabling an easier accommodation of substrates at low temperature. The fundamental issues concerning molecular basis of cold activity and the interplay between flexibility and catalytic efficiency are of importance in the study of structure-function relationships in enzymes. Such issues are often approached through comparison with the mesophilic or thermophilic counterparts, by site directed mutagenesis and three-dimensional crystal structures (Narinx et al. 1997; Wintrode et al. 2000; Papaleo et al. 2008). Therefore, 3D structural comparisons between mesophilic and psychrophilic protease Protease from psychrotrophic microorganisms have pH optima under alkaline conditions, but some of them have pH optima under neutral to acidic conditions. Thermo and alkalostability (stability under alkaline conditions) of protease is considered to be another important character, which is a useful criterion for application as a detergent additive due to the requirement of retention of activity during storage.

Economic Importance of Cold Adapted Proteases

Proteases are being used in the food and detergent industries since long time, but their application in leather industry and for organic waste treatment is a relatively new development with added biotechnological importance. Proteases are the most widely used enzymes in the detergent industry because of the inefficiency of nonenzymatic detergents to remove proteins from textile fibers. Cold active proteases with great catalytic efficiency at low temperature allow washing of textiles at room temperature (Demirjian et al. 2001). Recently there has been growing interest in psychrophilic enzymes as models in basic studies both to investigate the thermal stability of proteins and to understand the relationship between their stability, flexibility, catalytic efficiency, and as potential candidates for industrial and biotechnological applications (Gerday et al. 2000).

Many psychrotrophic microorganisms produce protease, though interest is limited only to those that yield substantial amounts of enzyme with desired properties. For example protease for detergent additive should be active in presence of oxidants and detergents, at low temperature, and stable under alkaline conditions. Here two things are of great importance: (1) selection of strain and (2) microorganism should be provided with optimal conditions to have good yield. It is not always true that the conditions which favor maximum growth will give highest enzyme production. For industrial scale production of proteases, technical media were usually employed which contained very high concentrations (100–150 g dry weight/liter) of complex carbohydrates, proteins, and other media components.

REFERENCES

Abo, M. *Method of purifying dry-cleaning solvent by decomposing liquid contaminants with a lipase.* World Organization Patent 9007606,1990.

Aghajari, N. et al. Crystallization and preliminary X-ray diffraction studies of a-amylase from the Antarctic psychrophiles *Alteromonashaloplanctis* A23. *Protein Sci.* 1996: 5, 2128–2129.

Alquati C., de Gioia L., Santarossa G., Alberghina L., Fantucci P., Lotti M. The cold-active lipase of Pseudomonas fragi: heterologous expression, biochemical characterization and molecular modeling. *Eur J Biochem* 2002; 269:3321–8.

Alvarez M., Zeelen J. P., Mainfroid V., Rentier-Delrue F., Martial J. A., et al. *J. Biol. Chem.* 1998;273:2199–206.

Atalla, R. H., and D. L. Vanderhart. Native cellulose: a composite of two distinct crystalline forms. *Science* 1984;223:283–285.

Babu Joseph, Pramod W. Ramteke, George Thomas: Cold active microbial lipases: Some hot issues and recent developments. *Biotechnology Advances* 26 2008; 457–470.

Bailey J. E., Ollis D. F. Applied enzyme catalysis. *Biochemical Engineering fundamentals.* New York: McGraw-Hill; 1986; 157–227.

Baldwin J., Hochachka P. W. *Biochem. J.* 1970116:883–87.

Barrett, J. Thermal hysteresis proteins. *Int. J. Biochem. Cell Biol.* 2001; 33, 105–117.

Beguin, P. and Anbert, J.P. The biological degradation of cellulose. *FEMS Microbiol. Rev.,* 1993; 13, 25-58.

Berger, F., Morellet, N., Menu, F. and Potier, P. Cold shock and cold acclimation proteins in the psychrotrophic bacterium Arthrobacter globiformis SI55. *J. Bacteriol.* 1996; 178, 2999–3007.

Bhatia R.P. *Contact lens cleaning composition containing an enzyme and a carboxylvinyl polymer*. United States Patent 1990;4921630.

Brocca S, Secundo F, Ossola M, Alberghina L, Carrea G, Lotti M. Sequence of the lid affects activity and specificity of C. rugosa lipase isozymes. *Protein Sci* 2003; 12:2312–9.

Brown, A.J., Ogawa, K. and Wood, T.M. Studies on the preparation and regeneration of protoplasts from the cellulolytic fungus, *Enzyme Micro & Technol,* 1986; 9, 527-532.

Brown, R. M., Jr., and I. M. Saxena. Cellulose biosynthesis: a model for understanding the assembly of biopolymers. *Plant Physiol. Biochem.* 2000;38**:**57–67.

Carrard, G., Koivula, A., Soderlund, H. and Beguin, P. Cellulose-binding domains promote hydrolysis of different sites on crystalline cellulose. *Proc. Natl. Acad. Sci. U.S.A.* 2000; 97, 10342–10347.

Cavicchioli, R., Thomas, T. and Curmi, P.M. Cold stress response in Archaea. *Extremophiles* 2000; 4, 321–331.

Chattopadhyay M. K. Mechanism of bacterial adaptation to low temperature. *J Biosci.* 2006; 31(1):157-65.

Chen I. C., Lin W. D., Hsu S. K., Thiruvengadam V., Hsu W. H. Isolation and characterization of a novel lysine racemase from a soil metagenomic library. *Appl Environ Microbiol* 2009; 75:5161–5166.

Cheng, C.H. Evolution of the diverse antifreeze proteins. *Curr. Opin. Genet. Dev.* 1998; 8, 715–720.

Chung E. J., Lim H. K., Kim J. C., Choi G. J., Park E. J., Lee M. H., Chung Y. R., Lee S. W. Forest soil metagenome gene cluster involved in antifungal activity expression in Escherichia coli. *Appl Environ Microbiol* 2008;74:723–730.

Coughlan, M.P. Cellulases: Production properties and applications. *Biochem. Soc. Trans.,* 1985;13, 405-406

Crawford D. L., Powers D. A. *Mol. Biol. Evol.* 1992;9:806–13.

D'Amico, S., Claverie P., Collins, T., Georlette D., Gratia, E., Hoyoux, A., Meuwis, M-A, Feller, G. and Gerday, C.. Molecular basis of cold adaptation. *Phil. Trans. R. Soc. Lond.* B 2002; 357, 917–925.

de Backer, M., McSweeney, S., Rasmussen, H.B., Riise, B.W., Lindley, P. and Hough, E. The 1.9 A crystal structure of heat-labile shrimp alkaline phosphatase. *J. Mol. Biol.* 2002;318, 1265– 1274.

Deming, J.W. Psychrophiles and polar regions. *Curr. Opin. Microbiol.* 2002; 5, 301–09.

Demirjian, D. C., Moris-Varas, F. & Cassidy, C. S. Enzymes from extremophiles. *Curr. Opin. Chem. Biol*2001;5, 144–151.

Devos N., Ingouff M., Loppes R., Matagne R. F. *J. Phycol.* 1998;34:655–60.

DeVries, A.L. (1988) The role of antifreeze glycopeptides and peptides in the freezing avoidance of Antarctic fishes. *Comp. Biochem. Physiol. B* 90, 611–621.

Ermolenko, D.N. and Makhatadze, G.I. Bacterial coldshock proteins. *Cell Mol. Life Sci.* 2002; 59, 1902–1913.

Ewart, K.V., Lin, Q. and Hew, C.L. Structure, function and evolution of antifreeze proteins. *Cell. Mol. Life Sci.* 1999;55, 271– 283.

Feller, G. & Gerday, C. Psychrophilic enzymes: molecular basis of cold adaptation. *Cell Mol. Life Sci.* 1997; 53, 830–841.

Feller, G., Arpigny, J. L., Narinx, E. & Gerday, C. Molecular adaptations of enzymes from psychrophilic organisms. *Comp. Biochem. Physiol.* 1997;118, 495–499.

Fields, M. W., J. B. Russell, and D. B. Wilson. 1998. The role of ruminal carboxymethylcellulases in the degradation of β-glucans from cereal grain. *FEMS Microbiol. Ecol.* 27:261–268.

Fujita J. Cold shock response in mammalian cells. *J. Mol. Microbiol. Biotechnol.* 1999;1:243–255.

Garsoux, G., Lamotte, J., Gerday, C., Feller, G., Kinetic and structural optimization to catalysis at low temperatures in a psychrophilic cellulase from the Antarctic bacterium *Pseudo alteromonashaloplanktis. Biochem. J.,* 2004;384, 247–253.

Georlette D., Blaise V., Collins T., D'Amico S., Gratia E., Hoyoux A., et al. Some like it cold: biocatalysis at low temperatures. *FEMS Microbiol Rev* 2004;28:25–42.

Gerday C., Aittaleb M., Bentahir M., Chessa J. P., Claverie P., Collins T., D'Amico S., Dumont J., Garsoux G., Georlette D., Hoyoux A., Lonhienne T., Meuwis M. A. and Feller G. Cold-adapted enzymes. from fundamentals to biotechnology; *Trends Biotechnol.* 2000;18, 103–107.

Gerday, C., Aittaleb, M., Arpigny, J. L., Baise, E., Chessa, J. P., Garsoux, G., Petrescu, I. & Feller, G. Psychrophilic enzymes: a thermodynamic challenge. *Biochim. Biophys. Acta* 1997;1342, 119–131.

Gianese G., Bossa F., Pascarella S. Comparative structural analysis of psychrophilic and meso- and thermophilic enzymes. *Proteins* 2002;47:236–49.

Goldstein, J., Pollitt, N. S. and Inouye, M. Major cold shock protein of *Escherichia coli*. *Proc. Natl. Acad. Sci. U.S.A.* 1990;87, 283-287.

Gomes J., Steiner W. The biocatalytic potential of extremophiles and extremozymes. *Food Technol Biotechnol* 2004;42:223–35.

Gotor-Fernandez V., Brieva R., Gotor V. Lipases: useful biocatalysts for the preparation of pharmaceuticals. *J Mol Catal B Enzym* 2006;40:111–20.

Graumann, P. and Marahiel, M.A. Some like it cold: response of microorganisms to cold shock. *Arch. Microbiol.* 1996; 166, 293–300.

Gualerzi C.O., Giuliodori A.M., Pon C.L. Transcriptional and post-transcriptional control of cold-shock genes. *J. Mol. Biol.* 2003;331:527–539.

Gupta R., Rathi P., BradoGupta R., Rathi P., Bradoo So S. Lipase mediated upgradation of dietary fats and oils. *Crit Rev Food Sci Nutr.* 2003;43(6):635-44.

Harding, M. M., Ward, L. G. and Haymet, A. D. Type I _antifreeze_ proteins. Structure–activity studies and mechanisms of ice growth inhibition. *Eur. J. Biochem.* 1999;264, 653–665.

Hayashi, N., J. Sugiyama, T. Okano, and M. Ishihara. The enzymatic susceptibility of cellulose microfibrils of the algal-bacterial type and the cotton-ramie type. *Carbohydr. Res.* 1997;305**:** 261–269.

Holmquist M. Alpha/beta-hydrolase folds enzymes: structures, functions and mechanisms. *Curr Prot Peptide Sci* 2000; 1:209–35.

Hu Y., Zhang G., Li A., Chen J., Ma L Cloning and enzymatic characterization of a xylanase gene from a soil-derived metagenomic library with an efficient approach. *Appl Microbiol Biotechnol* 2008;80:823–830.

Jaenicke, R. Protein stability and molecular adaptation to extreme conditions. *Eur. J. Biochem.* 1991;202, 715–728.

Jaenicke, R., Schurig, H., Beaucamp, N. and Ostendorp, R., Structure and stability of hyperstable proteins: glycolytic enzymes from hyperthermophilic bacterium Thermotogamaritima. *Adv. Protein Chem.* 1996;48, 181–269.

Jagdale G. B., Gordon R. *Comp. Biochem. Physiol A.* 1997;118:1151–56.

Jensen R. G., Hamosh M. Selectivity of lipases - types and determination. *Nato Adv Sci Inst Ser* 1969; 317:17–29.

Jia, Z. and Davies, P.L. Antifreeze proteins: an unusual receptor–ligand interaction. *Trends Biochem. Sci.* 2002; 27, 101–106.

Jones, P.G. and Inouye, M. The cold-shock response – a hot topic. *Mol. Microbiol.* 1994; 11, 811–818.

Kademi A., Danielle L., Ajain H. Lipases. In: Pandey A., Webb C., Soccol C. R., Larroche C., editors. *Enzyme Technology*. India: Asiatech Publishers; 2005; 297–318.

Kim, S.Y., Hwang, K.Y., Kim, S.H., Sung, H.C., Han, Y.S. and Cho, Y.J. Structural basis for cold adaptation. Sequence, biochemical properties, and crystal structure of malate dehydrogenase from a psychrophile Aquaspirilliumarcticum. *J. Biol. Chem.* 1999;274, 11761–11767.

Klibanov A. M. Improving enzymes by using them in organic solvents. *Nature* 2000; 1409:241–246.

Kobayashi, H., *Liquid leather cleaners,* Japanese Patent 1989;1225700,

Krishna H. S., Karanth N. G. Lipases and lipase-catalyzed esterification reactions in nonaqueous media. *Catal Rev* 2002;44:499–591.

Kulakova L., Galkin A., Nakayama T., Nishino T., Esaki N.Cold-active esterase from Psychrobacter sp. Ant300: gene cloning, characterization, and the effects of Gly Pro substitution near the active site on its catalytic activity and stability. *Biochim Biophys Acta*. 2004;1696(1):59-65.

Kumar S., Nussinov R. Different roles of electrostatics in heat and in cold: adaptation by citrate synthase. *Chem Bio Chem* 2004;5:280–90.

Kumar S., Tsai C. J., Nussinov R. Maximal stabilities of reversible two-state proteins. *Biochemistry* 2002 41:5359–74.

Lang D. A., Dijkstra B. W. Structural investigations of the regio- and enantioselectivity of lipases. *Chem Phys Lipids* 1998; 93:115–122.

LeCleir G. R., Buchan A., Maurer J., Moran M. A., Hollibaugh J. T. Comparison of chitinolytic enzymes from an alkaline, hypersaline lake and an estuary. *Environ Microbiol* 2007; 9:197–205.

Lee, S. J., Xie, A., Jiang, W., Etchegaray, J. P., Jones, P. G. and Inouye, M. Family of the major cold-shock protein, CspA (CS7.4), of Escherichia coli, whose members show a high sequence similarity with the eukaryotic Y-box binding proteins. *Mol. Microbiol* 1994;11, 833-839.

Leinala, E.K., Davies, P.L. and Jia, Z. Crystal structure ofb-helical antifreeze protein points to a general ice binding model. *Structure* 2002;10, 619–627.

M. Luisa Tutino, Guido di Prisco, Gennaro Marino and Donatella de Pascale. Cold-Adapted Esterases and Lipases: From Fundamentals to Application. *Protein & Peptide Letters*, 2009; 16, 1172-1180.

Makhatadze, G.I. and Privalov, P.L. Energetics of protein structure. *Adv. Protein Chem.* 1995; 47, 307–425.

Mandels, M., Applications of cellulases. *Biochem. Soc. Trans.,* 1985; 13, 414-415.

Margesin, R. Alpine microorganisms: useful tools for low temperature bioremediation. *J. Microbiol.,* 2007;*45*(4), 281-285.

Margesin, R., Feller, G., Gerday, C. and Russell, N. Cold-adapted microorganisms: adaptation strategies and biotechnological potential. In: *The Encyclopedia of Environmental Microbiology* (Bitton, G., Ed.), 2002; pp. 871–885. John Wiley & Sons, New York.

Mayr, B., Kaplan, T., Lechner, S. and Scherer, S. Identification and purification of a family of dimeric major cold shock protein homologs from the psychrotrophic Bacillus cereus WSBC 10201. *J. Bacteriol.* 1996; 178, 2916–2925.

Moriguchi, H., Hirata, J., Watanabe, T., *Microorganism based agent for treatment of organic wastes.* Japanese Patent 1990; 2105899.

Nakamura, K., Nasu, T. *Enzyme containing bleaching composition*, Japanese Patent.

Nardini M., Dijkstra B. W. α/β hydrolase fold enzymes: the family keeps growing. *Curr. opinstruct boil.* 1999;9:732–7.

Narinx E., Baise E., Gerday C. Subtilisin from psychrophilic Antarctic bacteria: characterization and site directed mutagenesis of residues possible involved in the adaptation to cold. *Protein Eng* 1997;10:1271–9.

Nushin Aghajari, Georges Feller, Charles Gerday, and Richard Haser'. Crystal structures of the psychrophilic a-amylase from *Alteromonas haloplanctis* in its native form and complexed with an inhibitor. *Protein Science* 1998; 7564-572.

Ollis D. L., Cheah E., Cygler M., Dijkstra B., Frolow F., Franken S. M., et al. The α/β hydrolase fold. *Protein Eng* 1992; 5:197–211. *Organic wastes.* Japanese Patent 1990;2105899,

Papaleo, E., Pasi, M., Riccardi, L., Sambi, I., Fantucci, P., De Gioia, L., Protein flexibility in psychrophilic and mesophilictrypsins. Evidence of evolutionary conservation of protein dynamics in trypsin-like serine-proteases. *FEBS Lett.,* 2008;582, 1008–1018.

Phadtare, S., Alsina, J. and Inouye, M. Cold-shock response and cold-shock proteins. *Curr. Opin. Microbiol.* 1999; 2, 175– 180.

Pinophilum P., Davies, G. and Henrissat, B., Structure and mechanisms of glycosyl hydrolases, *Struct.,* 1995; 3, 853-859.

Pleiss J, Fisher M, Schmid R. D. Anatomy of lipase binding sites: the sessile fatty acid binding site. *Chem Phys lipids* 1998;93:67–80 (Sp. Issue).

Privalov, P.L. Cold denaturation of proteins. *Crit. Rev. Biochem. Mol. Biol.* 1990;25, 281–305.

Rhee J. K., Ahn D. G., Kim Y. G., Oh J. W., New thermophilic and thermostable esterase with sequence similarity to the hormonesensitivelipase family, cloned from a metagenomic library. *Appl Environ Microbiol* 2005;71:817–825.

Rotticci D., Ottosson J., Norin T., Hult K. Candida antarctica lipase B: a tool for the preparation of optically active alcohols. *Methods Biotechnol* 2001;15:261–76.

Russell, N. J. Molecular adaptations in psychrophilic bacteria: potential for biotechnological applications. *Adv. Biochem. Engng Biotechnol* 1998;61, 1–21.

Russell, N. J. Toward a molecular understanding of cold activity of enzymes from psychrophiles. *Extremophiles* 2000;4, 83– 90.

Russell, R. J., Gerike, U., Danson, M. J., Hough, D. W. and Taylor, G. L. Structural adaptations of the cold-active citrate synthase from an Antarctic bacterium. *Structure* 1998; 6, 351– 361.

Ryu H. S., Kim H. K., Choi W. C., Kim M. H., Park S. Y., Han N. S., et al. Newcold-adapted lipase from Photobacteriumlipolyticum sp. nov.that is closely related to filamentous fungal lipases. *Appl Microbiol Biotechnol* 2006;70:321–6.

Santarossa G., Lafranconi P. G., Claudia A., Luca D. G., Lilia A., Piercarlo F., et al. Mutations in the "lid" region affect chain length specificity and thermostability of a Pseudomonas fragi lipase. *FEBS Lett* 2005;579:2383–6.

Saunders N. F., Thomas T., Curmi P. M., Mattick J. S., Kuczek E., et al. *Genome Res.* 2003;13:1580–88.

Schofield, P., R. E. Pitt, and A. N. Pell. 1994. Kinetics of fiber digestion from in vitro gas production. *J. Anim. Sci.* 72:2980–2991.

Sharma S., Khan F. G., Qazi G. N. Molecular cloning and characterization of amylase from soil metagenomic library derived from Northwestern Himalayas. *Appl Microbiol Biotechnol* 2010; 86:1821–1828.

Shoham, Y., Lamed, R. and Bayer, E. A. The cellulosome concept as an efficientmicrobial strategy for the degradation of insoluble polysaccharides. *Trends Microbiol.* 1999; **7**, 275–281.

Siddiqui, K.S., Cavicchioli, R. Cold-adapted enzymes. *Annu. Rev. Biochem.*, 2006, *75*, 403- 433.

Singh A, Singh Chauhan N, Thulasiram HV, Taneja V, Sharma R., Identification of two flavinmonooxygenases from an effluent treatment plant sludge metagenomic library. *Bioresour Technol,* 2006, 101.8481–8484.

Smalas, A. O., Leiros, H. K., Os, V. Cold adapted enzymes. *Biotechnol. A. Rev. 2000;*6, 1–57.

Smalas, A.O., Heimstad, E.S., Hordvik, A., Willassen, N.P. and Male, R. Cold adaption of enzymes: structural comparison between salmon and bovine trypsins. *Proteins* 1994;20, 149–166.

Somero G. N. *Annu. Rev. Physiol* 1995;57:43–68.

Tan S., Owusu A. R. K., Knapp J. Low temperature organic phase biocatalysis using cold adapted lipase from psychrotrophic Pseudomonas P38. *Food Chem* 1996;57:415–8.

Tomme, P., Warren, R. A. and Gilkes, N. R. Cellulose hydrolysis by bacteria and fungi. *Adv. Microb. Physiol.* 1995; 37, 1–81.

Toyota, E., Ng, K.K., Kuninaga, S., Sekizaki, H., Itoh, K., Tanizawa, K. and James, M.N. *Crystal structure and nucleotide sequence of an anionic trypsin from chum salmon* 2002.

Uppenberg J., Hansen M. T., Patkar S., Jones T. A. The sequence, crystal structure determination and refinement of two crystal forms of lipase B from Candida antarctica. *Structure* 1994b;2:293–308.

Uppenberg J., Patkar S., Bergfors T., Jones T. A. Crystallization and preliminary X-ray tudies of lipase B from Candida antarctica. *J Mol Biol* 1994a;235:790–2.

Verdier, J.M., Ewart, K.V., Griffith, M. and Hew, C.L. An immune response to ice crystals in North Atlantic fishes. *Eur. J. Biochem.* 1996; 241, 740–743.

Verger R. Enzyme kinetics of lipolysis. *Methods Enzymol* 1980;64:341–92.

Vieille, C. and Zeikus, G.J. Hyperthermophilic enzymes: sources, uses, and molecular mechanisms for thermostability. *Microbiol. Mol. Biol.Rev.* 2001; 65, 1–43.

Violot S., Aghajari N., Czjzek M., Feller G., Sonan G. K., et al. *J. Mol. Biol.* 2005;348:1211–

Waschkowitz T., Rockstroh S., Daniel R. Isolation and characterization of metalloproteases with a novel domain structure by construction and screening of metagenomic libraries. *Appl Environ Microbiol* 2009; 75:2506–2516.

Weimer, P. J., J. M. Hackney, H.-J. G. Jung, and R. D. Hatfield**.** 2000. Fermentation of a bacterial cellulose/xylan composite by mixed ruminalmicroflora: implications for the role of polysaccharide matrix interactions in plant cell wall biodegradability. *J. Agric. Food Chem.* 48:1727–1733.

Wintrode P. L., Miyazaki K., Arnold F. H. Cold adaptation of a mesophilicsubtilisin-like protease by laboratory evolution. *J Biol Chem* 2000; 275:31635–40.with a lipase. World Organization Patent 9007606, 1990.

Yamanaka, K., Fang, L. and Inouye, M. The CspA family in *Escherichia coli*: multiple gene duplication for stress adaptation. *Mol. Microbiol.* 1998; 27, 247-255.

Ye M., Li G., Liang W. Q., Liu YH Molecular cloning and characterization of a novel metagenome-derived multicopper oxidase with alkaline laccase activity and highly soluble expression. *Appl Microbiol Biotechnol* 2010; 87:1023–1031.

Zavodszky, P., Kardos, J., Svingor, A. & Petsko, G. A. Adjustment of conformational flexibility is a key event in the thermal adaptation of proteins. *Proc. Natl Acad. Sci. USA* 1998;95, 7406–7411.

Zecchinon, L., Paule Claverie Tony Collins, Salvino D'Amico, Daniel Delille, Georges Feller, Daphné Georlette, Emmanuelle Gratia, Anne Hoyoux, Marie-Alice Meuwis, Guillaume Sonan, Charles Gerday. Did psychrophilic enzymes really win the challenge? *Extremophiles* 2001; 5:313–321.

In: Recent Advances in Microbiology. Volume 3 ISBN: 978-1-53614-057-6
Editors: N.C. Gautam, S.P. Tiwari and R. Sharma

Chapter 7

PRODUCTION AND POTENTIAL APPLICATIONS OF SINGLE CELL PROTEIN

***D. V. Singh*[1,*]*, P. Pathak*[1]*, P. Malviya*[1] *and U. P. Singh*[2]**
[1]Department of Botany, Udai Pratap College (Autonomous),
MGK Vidyapith, Varanasi UP, India
[2]Amity Institute of Biotechnology,
Amity University Uttar Pradesh, UP, India

INTRODUCTION

Single cell protein (SCP) refers to dried microbial cells or total protein extracted from pure microbial biomass (algae, bacteria, filamentous fungi and yeasts), which can be used as human food (food grade) or animal feed additive or supplement (Anupama and Ravindra, 2000). Due to rapid growth in human population, deficiency of nutrients, especially protein, has been observed in human as well as animal diet. Single cell protein have multiple applications in animal nutrition, such as fattening of calves, poultry, pigs and fish breading. In food, SCP is used as aroma carrier, vitamin carrier and emulsifying aid, apart from improving the nutritive value of baked products in ready-to-serve meals in diet recipes. Also, SCP has found usage in the manufacturing sector, in paper processing, leather processing and as foam stabilizers.

The term single cell protein (SCP) was coined in 1966 by Carol L. Wilson at the Massachusetts Institute of Technology (MIT), U.S.A. Rising populations and the need for relieving protein malnutrition in developing countries has catalyzed several efforts to

---

* Corresponding author: drdurgvijay@gmail.com

intensify the production of protein both from conventional agriculture as well as non-conventional sources. Microbial biomass/cells or cell cultures like algae (*Spirulina*), yeasts/fungi (*Saccharomyces cerevisiae*) and bacteria (*Methylophilus methylotrophus*) as a source of SCP has attracted the attention of biologists and technologists as promising options to reduce the protein demand-supply gap. The greatest promise of SCP, particularly in India, lies in its potential as animal and poultry feed and also as protein supplement for our largely vegetarian consumers (Kharatyan, 1978).

It has been estimated that if necessary measures are not taken, the malnutrition problem will lead to some major health crises in the developing countries. Therefore, it is very important to increase protein production and also make it available to the general population by utilizing all the available methods. The increasing global demands for food and feed protein spurred the search for non-conventional protein sources (Nasseri et al., 2011). A great deal of research interest has been focused on the potential of converting agricultural wastes to microbial protein.

Keywords: single cell protein, SCP, spirullina, fungi, algae

History

Research on single cell protein technology started a century ago when Max Delbruck and his colleagues estimated the high value of surplus Brewer's yeast as a feeding supplement for animals. During World War I, SCP technology proved to be more than useful as Germany used it to replace more than half of its imported protein sources by yeast.

The first industrial production of SCP was done in the form of 'Torula' yeast (*Candida utilis*) in Germany and was used in soups and sausages. Even earlier than that, ancient populations of the Aztecs near Lake Texcoco, Mexico and the Kanem tribe near Lake Chad, Central Africa have been consuming sun-dried *Spirulina*, the filamentous blue-green alga harvested from the lake.

Recently in many countries *Spirulina* has attained the status of a health food item in view of its nutritional excellence (Cifferri, 1983).

Sources of SCP

Microorganisms like algae, fungi/yeasts and bacteria are the chief sources of single cell protein and they can be used for large-scale SCP production.

Algae

(a) As discussed above, *Spirulina* has been used as human food since ages.
(b) Algae are used as a food in many different ways and its advantages include effective utilization of solar energy, faster growth and richness in protein content (Becker and Venkataraman, 1982; Cifferri, 1983).
(c) *Spirulina maxima*, *Chlorella* and *Scenedesmus acutus* have been cultivated for SCP. Among the microalgae, the genus *Spirulina* contains nearly 60-70% of proteins, nucleic acids and amino acids recommended by the Food and Agriculture Organization (Pelizer et al., 2003).
(d) They contain about 60% protein by weight, with rich amino acid composition, but do lack in sulfur-containing amino acids.
(e) *Chorella* and *Spirulina* are used for commercial scale production in Taiwan, Thailand, Japan, Israel, Mexico, USA and India.
(f) It is spray-dried and sold as pills and powders.
(g) *Spirulina* grown on sewage water is free of pathogenic microbes.
(h) *Spirulina* is the most popularly consumed alga, with even astronauts bringing it to their space-crafts.

Disadvantages

(a) Most algal species (except *Spirulina*) are not suitable for human consumption because they are rich in chlorophyll.
(b) Algal biomass has low density, i.e., 1-2 gm dry weight/liter of substrate.
(c) There is a substantial risk of contamination during growth.

Yeasts and Fungi

(a) Many fungal species are used as a source of protein-rich food. Among these, most popular are yeast species *Candida*, *Hansenula*, *Pitchia*, *Torulosis* and *Saccharomyces.*
(b) Many filamentous fungi such as *Chaetomium cellulolyticum* grow on cellulose waste.
(c) *Fusarium graminearum* grows on starch and *Paecilomyces varioti* grows on sulfur liquor and both are used for the production of SCP. These have about 50-55% protein by weight.
(d) Yeasts such as *Candida utilis* (*Torula* Yeast) and *Candida lipolytica* growing on ethanol and *Saccharomyces cerevisiae* growing on molasses are used for SCP production.
(e) Several species of mushrooms are used as protein-rich food.

Table 2. The commonly used microorganisms for SCP production

Microorganism	Substrate
Bacteria	
Methylophilus methylotrophus	Methanol
Methylomonas methylotrophus, M. clara	Methanol
Aeromonas hydrophylla	Lactose
Acromobacter delvacvate	*n*-alkanes
Acinetobacter calcoacenticus	Ethanol
Bacillus megaterium	Non-protein nitrogenous compound
Bacillus subtilis	Cellulose, Hemicellulose
Cellulomonas sp.	Cellulose, Hemicellulose
Flavobacteriun sp.	Cellulose, Hemicellulose
Thermomonospora fusca	Cellulose, Hemicellulose
Lactobacillus sp.	Glucose, Amylose, Maltose
Pseudomonas fluorescens	Uric acid and other non protein nitrogenous compounds
Rhodopseudomonas capsulate	Glucose
Brevibacterium sp.	C_1-C_2 hydrocarbons
Fungi	
Aspergillus fumigates	Maltose, Glucose
Aspergillus niger, Aspergillus oryzae	Cellulose, Hemicellulose
Cephalosporium eichhorniae	Cellulose, Hemicellulose
Chaetomium cellulolyticum	Cellulose, Hemicellulose
Penicillium cyclopium	Glucose, Lactose, Galactose
Rhizopus chinensis	Glucose, Maltose
Scytalidium aciduphlium	Cellulose, Pentose
Trichoderma viridae, Trichoderma alba	Cellulose, Pentose
Fusarium graminearum	Starch
Yeasts	
Saccharomyces cerevisiae	Lactose, Pentose, Maltose, Molasses
Candida tropicalis	Maltose, Glucose
Candida utilis	Glucose
Candida novellas	*n*-alkanes
Candida intermedia	Lactose
Amoco torula	Ethanol
Algae	
Spirulina maxima	Carbon dioxide through photosynthesis
Chondrus crispus	Carbon dioxide through photosynthesis
Scenedesmus sp.	Carbon dioxide through photosynthesis
Chlorella pyrenoidosa, C. soroleiana	Carbon dioxide through photosynthesis
Porphyrium sp.	Carbon dioxide through photosynthesis

SUBSTRATES FOR SCP PRODUCTION

The potential substrates/raw materials that can be used for single cell protein production include whey, cellulose, sulfite waste liquor, hydrocarbon waste from petroleum industry and the vats used to produce alcoholic beverages (Chanda and Sibani, 1996; Ugalde and Castrillo, 2002). Single cell proteins develop when microbes ferment waste materials (including food processing wastes, straw, cannery, wood, residues from alcohol production, hydrocarbons or human and animal excreta). The most widespread and commonly used substrates for SCP production have been those where the carbon and energy source is derived from carbohydrates. This is due to the fact that their building blocks are natural microbial substrates and that carbohydrates are a renewable resource that is widely distributed. The choice of substrates that are normally abundant and in proximity to the production plant has determined the design and strategy of SCP production processes.

A variety of substrates (Table 3) ranging from inorganic carbon, e.g., CO_2 (no cost for substrate), through industrial effluents, e.g., confectionery effluents, whey, etc. (utilization of the substrate helps effluent disposal) and low cost organic materials, e.g., cellulosic wastes like straw, etc., to high cost materials like starch hydrolysate are used for SCP production. These substrates can be divided into two broad groups: (1) renewable carbon sources and (2) fossil carbon sources (non-renewable).

1. **Renewable carbon sources:** The renewable carbon sources include CO_2, molasses, whey, cellulose hydrolysates, starch hyrolysates, industrial effluents and cellulosic wastes. SCP production using these substrates is briefly described below:-
   (a) **CO_2:** This substrate is utilized by algae, which derive the required energy from sunlight. *Spirulina* is grown on a commercial scale by Sosa Texcoco Co., Mexico, producing up to 5 tons dry biomass per day. *Chlorella* is commercially grown in Japan and Taiwan to get about 15,000 tons/yr dry biomass from many small production units.
   (b) **Molasses:** This substrate is used for alcoholic fermentation from which yeast biomass is obtained as a byproduct. Traditionally baker's yeast (*Saccharomyces cerevisiae*) and Torula yeast (*Candida utilis*) are produced on this substrate. About 200,000 tons dry weight of *S. cerevisiae* is produced each year, which is used as inoculum for dough fermentation in bakeries. Four types of molasses are available: cane, beat, corn and citrus. Cane molasses has been used for SCP production using *Candida utilis* or *Saccharomyces cerevisiae* and it is produced in quite large volumes by a number of countries. Molasses may require pre-treatment to reduce mineral content and remove suspended solids before use.
   (c) **Whey:** Whey is the liquid portion of milk that remains after the curd is separated during cheese production. On dry weight basis, whey contains 70% lactose, 9-14% protein and 9% ash. The 'Bel' yeast process has been successful in using *Kluyveromyces fragilis* as the organism to produce SCP from lactose. Problems associated with whey for SCP production are usually

seasonal supply variation and its high water content, which makes transport prohibitively expensive.

(d) **Cellulose hydrolysate:** Cellulose obtained from fungi like *Trichoderma viridae* has been used to hydrolyze cellulose and to produce glucose. The main advantage of this approach is the easy availability of large quantities of cheap substrate. However, the challenges are (i) difficult and expensive process of complete hydrolysis of lignocellulose, (ii) the amount of pretreatment necessary and (iii) the extent of sugar destruction and by-product formation during chemical hydrolysis. The best utilization of cellulosic substrates appears to be for semisolid fermentation systems.

(e) **Starch hydrolysate:** This expensive substrate is used for SCP production using *Fusarium graminearum*. The SCP has a fibrous, meat-like texture and is marketed as 'Mycoprotein.' It is approved for human use in the U.K. and food products having Mycoprotein are being marketed.

(f) **Industrial effluents:** Effluents from many industries, e.g., breweries, distilleries, confectionery industries, potato and canning industries, sulfite liquor from wood pulp mills, etc., contain a large amount of carbohydrates and other organic compounds, which can be used for SCP production. *Candida utilis* is used to treat confectionery effluent. Sulfite liquor is widely used for SCP production in North America, Europe and the erstwhile Soviet Union. The most advanced process is based on the fungus *Paecilomyces varioti*, which is being used on an industrial scale in Finland. The fungus utilizes both pentose and hexose sugars as well as the acetic acid present in the liquor.

(g) **Cellulosic wastes:** A new approach is to use agricultural and forestry cellulosic wastes, e.g., straw, bagasse, sawdust, etc., for SCP production, as per the principle of semisolid fermentation. The cellulosic material is pretreated thermally and chemically and is then fermented with the fungus *Chaetomium cellulolyticum*. The fungus is reported to degrade the cellulosic wastes within a very short time, to the tune of 4 hrs.

2. Fossil carbon sources (Non-renewable):

The various substrates included in this group are gaseous hydrocarbons, liquid hydrocarbons, methanol and ethanol. These substrates are briefly described below:-

(a) **Gaseous hydrocarbons:** Gaseous hydrocarbons (C_1-C_4) have been used for SCP production. Methane was initially considered as an SCP raw material because as a gas product, purification problems after fermentation would be minimal. Methane is utilized by certain bacteria including *Pseudomonas methanica* and *Methanomonas methanica*. Methane is highly desirable as a substrate, as it is available in high purity from natural gas, which in some parts of the world is burnt in the oil wells, e.g., Bombay High. In addition, it is readily removed from the fermentation medium and supports high productivity in continuous processes. However, disadvantages associated with methane-based processes are related to (i) the greater oxygen requirement necessary to fully oxidize methane as compared to paraffins (ii) the low solubility of methane in water and (iii) the fermentation plant should be flame-proof as methane-oxygen mixtures are highly explosive.

(b) **Liquid hydrocarbons:** Saturated, straight-chain hydrocarbons called *n*-alkanes, constitute 0-30% of crude oil. C_5-C_8 *n*-alkanes are liquid at room temperature but are generally toxic to cells due to their solvent action properties. Generally, *n*-alkanes having 9 to 18 carbons (C_9-C_{18}) are used as SCP substrates. Many bacteria, actinomycetes, yeasts and molds are able to use liquid hydrocarbons. Gas oil, in combination with ammonia and mineral nutrients, was used in air-lift fermenters to produce *Candida* yeast by British Petroleum. Pure *n*-alkanes having 10 to 23 carbons were used by British Petroleum to culture *Candida lipolytica* in a continuous process; the medium consisted of *n*-alkanes, minerals and ammonia. The *n*-alkanes were almost completely utilized, biomass recovery was easier and cheaper and the SCP with about 60% crude protein was marketed as 'Toprina'. The SCP was comparable to soyabean meal and fishmeal and was cleared after nutritional and toxicological tests.

(c) **Methanol:** A very efficient chemical conversion of methane yields methanol; it can also be produced from coal, gas oil, wood, naphtha, etc. Methanol presents the dual advantages of a clean and reliable substrate giving a high yield of SCP together with easy availability and a reasonable price. The use of methanol as a substrate for SCP has been developed by ICI to full–scale commercial production. Methanol technology has great potential for areas of the world with reliable methane supplies and a need for an indigenous source of high quality protein. Methanol is a good substrate for the production of SCP using *Methylophilus methylotrophus*.

(d) **Ethanol:** Ethanol has many of the advantages of methanol, but is more expensive. It can be obtained from ethylene by catalytic addition of a water molecule or from organic substrates by alcoholic fermentation. It is more acceptable as a substrate for SCP production for human use. Several bacteria, yeasts and filamentous fungi utilize ethanol.

Commercial Production/Cultivation of Single Cell Protein

The SCP production process involves growth of the microorganisms in large fermentation tanks with forced aeration for vigorous cell growth. High biomass productivity requires high oxygen transfer rates, which promote high respiration, which in turn increases metabolic heat production, thus justifying the need of an efficient cooling system. In order to maximize fermentation productivity, it is essential to go for continuous and not batch or fed-batch fermentation processes (Stanbury et al., 2000).

The process of SCP production from any microorganism/substrate would involve the following basic steps:

1. Preparation of suitable culture medium.
2. Provision of a carbon source, which may need physical and/or chemical pre-treatments.
3. Addition of sources of nitrogen and phosphorous and other nutrients needed to support optimal growth of selected microorganism.
4. Prevention of contamination by maintaining aseptic conditions. The medium components may be sterilized by filtration or heating. Fermentation equipment may also be sterilized.
5. Inoculation of the selected microorganism in the form of a pure culture.
6. Provision of adequate aeration as SCP processes are highly aerobic (except those using algae).
7. Recovery of microbial biomass from the medium.
8. Processing of the biomass for enhancing its usefulness.

Single cell protein basically comprises proteins, fats, carbohydrates, ash ingredients, water and other elements such as phosphorous and potassium. The composition depends upon the organism and the substrate on which it grows. If SCP is to be used successfully, five main criteria need to be satisfied:

1. The SCP must be safe to eat.
2. The nutritional value dependent on the amino acid composition must be high.
3. It must be acceptable to the general public.
4. It must have the functionality, i.e., characteristics, which are found in common staple foods.

5. The economic viability of the process should be demonstrable. Single cell protein production process involves two types of fermentation processes namely:

- Submerged fermentation
- Semisolid state fermentation

Table 3. Various potential substrates for SCP production using different microorganism

Substrate	Microorganism
A. Renewable Carbon Sources	
1. Carbon dioxide (CO_2)	*Spirulina, Chlorella*
2. Molasses	*Saccharomyces cerevisiae* (baker's yeast)
3. Whey	*Kluyveromyces fragilis, Lactobacillus bulgaricus, Candida krusei, Candida intermedia*
4. Starch hydrolysates	*Fusarium graminearum*
5. Industrial effluents (a) Confectionery effluents (b) Sulphite-liquor	 (a) *Candida utilis* (b) *Paecilomyces varioti, Candida utilis*
6. Cellulosic wastes, e.g., straw, sawdust, bagasse, etc.	*Chaetomium cellulolyticum*
B. Fossil Carbon Sources	
1. Liquid hydrocarbons	*Candida lipolytica*
2. Gaseous hydrocarbons (a) Methane (b) C_1-C_4 hydrocarbons	 (a) *Pseudomonas methanica, Methanomonas methanica* (b) *Brevibacterium sp.*
3. Methanol	Bacteria: *Methylophilus methylotrophus* Other Bacteria: *Methylomonas methanolica* Yeasts: *Torulopsis glabrata, Candida boidinii*, etc.
4. Ethanol	*Candida utilis*

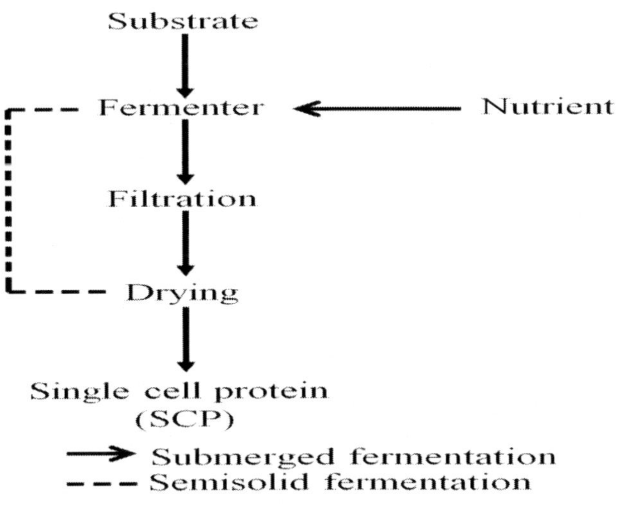

Figure 1. Flow diagram for cultivation of single cell protein.

Submerged Fermentation

The submerged process always requires the substrate to be in the liquid state, which dissolves the nutrients needed for growth. The fermenter that contains the substrate is operated continuously and the product biomass is continuously harvested from the fermenter by using different techniques. Then, the product is filtered or centrifuged and dried.

Semisolid Fermentation

Semisolid fermentation generally requires substrates in the solid state, e.g., cassava waste. Submerged culture fermentations require more capital investment and have higher operating cost (Nasseri et al., 2011; Ritzka et al., 1997).

Table 4. Chemical composition and nutritional value of SCP from the respective organisms

Component	Percentage composition by weight		
	Algae	Fungi/Yeasts	Bacteria
True protein	40 - 60[a]	30 - 70[a]	50 - 83[a]
Total nitrogen (protein + nucleic acids)	45 - 65[a]	35 - 50[a]	60 - 80[a]
Lysin	4.6 - 7.0[a]	6.5 - 7.8[a]	4.3 - 5.8[a]
Methionine	1.4 - 2.6[a]	1.5 - 1.8[a]	2.2 - 3.0[a]
Fats/Lipids	5-10[a]	5 - 13[a]	8 - 10[a]
Carbohydrates	9	NA	NA
Bile pigments and Chlorophyll	6	NA	NA
Nucleic acids	4 - 6[a]	7 - 10	15 - 16[a]
Mineral acids	7	6.6	8.6
Amino acids	NA	54	65
Ash	3	NA	NA
Moisture	6.0	4.5 - 60[a]	2.8
Fiber	3	NA	NA

[a]The yield varies with type of substrate used, the specific organism used and the culture condition maintained; NA-Not available.

The cultivation involves many operations, which includes stirring and mixing of a multiphase system, transport of oxygen from the gas bubbles through the liquid phase to the microorganisms and the process of heat transfer from liquid phase to the surroundings. A special bioreactor is designed to intensify the mass and energy transportation phenomena called U-loop fermenter.

Chemical Composition and Nutritional Value

Chemical composition of the SCP biomass is the most important factor to be evaluated before finalizing its potential for utilization as feed or food. To assess the nutritional value of SCP, many factors must be considered, which include nutrient composition, amino acid profile, vitamin and nucleic acid content as well as allergies and gastrointestinal effects. Single cell protein is normally considered as a source of protein. However, like any other biological material, it also contains nucleic acids, carbohydrate, cell wall materials, lipids, minerals and vitamins (Bhalla et al., 2007). To assess toxicological and carcinogenic effects, long-term feeding trials are also required. In addition, the processes of drying, harvesting and processing have an effect on the nutritive values of the finished product. Table 4 shows the comparison of the compositions of SCP from the different species commercially used for the purpose.

The yield varies with type of substrate used, the specific organism used and the culture condition maintained. The parameters used in the table evaluate the ranking of SCP, showing that nucleic acids from algae is safer than those obtained from fungi and bacteria. Nucleic acids from fungi is comparatively better from those we get from bacteria due to moderate levels of the same. Therefore, the ranking will be Algae > Fungi > Bacteria.

Single Cell Protein Production: Product Safety and Quality

SCP has applications in animal feed, human food and as functional protein concentrations. Some contaminants can produce mycotoxins. Some bacterial SCP have amino acid profiles similar to animal/plants protein. Yeast, fungal and soybean proteins tend to be deficient in methionine. Microbial biomass has high nucleic acid content, and levels must be limited in the diets of monogastric animals to <50 gm per day. Ingestion of purine compounds arising from RNA breakdown leads to increased plasma levels of uric acid, which can cause gout and kidney stones. High content of nucleic acids causes no problems to animals since uric acid is converted to allantoin, which is readily excreted in urine. Hence, nucleic acid removal is not necessary from animal feeds but it is important to do so from human foods. A temperature hold at 64°C reduces intracellular RNA levels in *Fusarium graminearum* from 80 mg/g to 2mg/g. Hence, the presence and quantity of possible toxic or carcinogenic compounds must be assayed for. These compounds may have been present in the substrate, may be synthesized by the organism or produced during the processing of SCP. Microbiological description, e.g., species and strain, should be provided and information on contamination be also given. The nutritional value should be evaluated on the target species and other species should also be included. The products for human use should necessarily be evaluated over a longer period, using a multistage process.

Advantages of SCP

Single cell protein (dried microbial biomass) as a supplement to human food or animal feed has the following advantages:-

1. It has very high protein content (40-70%) in the dry mass.
2. It has a very low fat content.
3. It is a good source of vitamins particularly B-complex, e.g., Yeasts.
4. It has a very high rate of multiplication. Microorganisms used as SCP grow faster, having a very shorter generation time (Algae 2-6 hrs, Yeasts 1-3 hrs and Bacteria 1/2-2 hrs).
5. They have lower amount of nucleic acids (3-4%).
6. They are easy to grow and not dependent of climatic variations. Microbial biomass production can occur in continuous cultures.
7. During the production of SCP biomass some organisms produce useful by-products such as organic acids and fats.
8. Most of the essential amino acids are present in single cell proteins.
9. They use low cost substrates and in some cases, such substrates that are being wasted and causing pollution to the environment. Hence, SCP production also helps reduce pollution.
10. At present, SCP appears to be the only feasible approach to bridge the gap between requirement and supply of protein.

Disadvantages

There are certain disadvantages associated with using SCP (dried microbial biomass) as supplement to human food/animal feed:-

1. Many microbes have high nucleic acid content, which can cause toxicity.
2. Poor digestibility and gastrointestinal problems in humans, since algae have cellulose in their cell walls, which are not digestible. Algae can also accumulate heavy metals, which may prove harmful to living beings (Becker, 2007).
3. SCP may cost more than conventional food products.

Economics of SCP Production

The spectacular development of fermentation technology in the last few decades has not taken place without sophisticated evolution of the economic viability of processes. In the case of SCP production, the need for accurate cost estimation is very relevant, since in the majority of cases the product is competing against protein sources of plant origin and the profit margins are predictably low.

The initial reasoning by companies such as British Petroleum and Imperial Chemical Industries to enter into SCP production was to produce, at low cost, high value SCP from petroleum for addition to animal feed. The intention was to replace imported protein additives such as soybean meal. Factors that contributed to the failure of hydrocarbon SCP to make a major commercial impact included the 1973 dramatic oil price increase, which raised feedstock and energy costs and the lower price increases achieved by agricultural products such as soybean. When one considers that crude oil prices increased by a factor of 6 in 1973 and that the cost of substrate for SCP processes represents 40-60% of total manufacturing costs, the negative impact on hydrocarbon-based SCP processes may be understood (Gibbs and Grrenhalgh, 1983).

Agricultural crops, the major competitor to SCP for animal feed, manifest a remarkable ability to respond to market forces and maintain prices stability. Hence, an added economic effort is required to promote the product against such an established competitor. By producing human protein supplement, the end value of product can be appreciated; however, the production process has to utilize more conventional and costly substrates for production (i.e., glucose). Thus, in all cases, product cost is a central element in the food market industry.

Production of SCP using cheap materials as substrate provides an economically feasible source of protein for use in animal feed or the processing of products for human consumption, as it often meets dietary requirements for protein. Many microorganisms have been used to convert various substrates into biomass. SCP production technologies arose as a promising way to solve the problem of worldwide protein shortage. They evolved as bioconversion processes, which turned low-value by-products into products with added nutritional and market value. Since SCP belongs to one of the cheapest protein products in the market, its production is profitable (Ugalde and Castrillo, 2002).

The economic factors that should be taken into account during the SCP production period are investment, energy, operating cost, waste, safety and the global market.

Substrate Costs

The substrate costs are the largest single cost factor. The manufacture of raw materials is more economical in larger plants. In the case of SCP production, the raw material accounts for 62% of the total product cost, followed by fixed charges attributed to the production process.

Utilities

The energy for compressing, air cooling, sterilizing and drying constitutes the next most important cost factors.

Capital Load

The capital load-dependent costs are determined by the cost of the apparatus for the process, the capacity of a plant and the capacity conditions. The main variable here is the size of the plant.

Product-Specific Variables

Process costs are dependent on the product characteristics, bulk and quality being the foremost.

REFERENCES

Anupama, R. & Ravindra, P. (2000). Value-added food: Single cell protein. *Biotechnology Advances*, *18*, 459-479.

Becker, E. W. (2007). Micro-algae as a source of protein. *Biotechnol. Adv.*, *25*, 207-210.

Becker, E. W. & Venkataraman, L. V. (1982). Biotechnology and exploitation of Algae-The Indian Approach. *German Agency for Technical Co-operation (GTZ) Eschborn (FRG)*, 1-216.

Bhalla, T. C., Sharma, N. N. & Sharma, M. (2007). *Production of Metabolites, Industrial Enzymes, Amino Acids, Organic Acids, Antibiotics, Vitamins and Single Cell Proteins.* National Science Digital Library, India.

Chanda, S. & Sibani, C. (1996). Plant origin liquid waste: A source for single cell protein production by Yeast. *Bioresource Technology*, *57* (1), 51-54.

Ciferri, O. (1983). *Spirulina*, the edible microorganism. *Microbiol. Rev.*, *47*, 551-578.

Gibbs, D. F. & Grrenhalgh, M. E. (1983). *Biotechnology, Chemical Feedstocks and Energy Utilization.* Frances Printer (Publ.), Washington.

Kharatyan, S. G. (1978). Microbes as Food for humans. *Ann. Rev. Microbiol*, *32*, 301-307.

Nasseri, A. T., Rasoul-Amini, S., Morowvat, M. H. & Ghasemi, Y. (2011). Single cell protein: Production and process. *Ame. J. Food Technol.*, *6*, 103-116.

Pelizer, L. H., Danesi, E. D. G. & Rangel, C. O. (2003). Influence of inoculums age and concentration in *Spirulina plantesis* cultivation. *J. Food Eng.*, *56*, 371-375.

Ritzka, A., Sosnitza, P., Ulber, R. & Sheper, T. (1997). Fermentation monitoring and process control. *Curr. Opin. Biotechnol.*, *8* (2), 160-164.

Stanbury, P. F., Whitaker, A. & Hall, S. J. (2000). *Principles of Fermentation Technology. 2nd Edn.*, Elsevier Applied Science Publishers, New York, USA.

Ugalde, U. O. & Castrillo, J. I. (2002). Single cell proteins from fungi and yeasts. *Applied Mycology and Biotechnology*, *2*, 123-149.

In: Recent Advances in Microbiology. Volume 3 ISBN: 978-1-53614-057-6
Editors: N.C. Gautam, S.P. Tiwari and R. Sharma

Chapter 8

INFRA-RED SPECTROSCOPY: CONCERNING THE SCOPE OF FOOD MICROBIOLOGY

Pranita Jaiswal**, Manpreet Kaur Grewal, Shyam Narayan Jha and Rishi Bhardwaj
Agricultural Structures and Environmental Control Division,
Central Institute of Postharvest Engineering and Technology, Ludhiana, India

INTRODUCTION

Food and Agriculture is the largest sector in the national economy. Before reaching the consumer, food passes through many hands and is handled and processed in several ways. To retain its quality, it is essential to ensure the integrity and safety of food throughout the supply chain. Measures taken to ensure the above will also add to the value of food as a commodity and lead to better acceptability from the consumer, both domestic and international. According to the reports released by the World Health Organization (WHO) and Food Agriculture Organization (FAO) of the United Nations, illness due to contaminated food is one of the most widespread health problems in the world and an important cause of reduced economic productivity, especially in developing and under-developed countries. Currently, millions of people worldwide are suffering from diseases caused by contaminated food, which inflicts a heavy social and economic cost. Major challenge of spoilage in the food industry is due to microbial contamination. While there is no easy solution to this problem, developing rapid tests for detection of food borne pathogens can be a significant contribution. Growth in consumer demand for safe food has increased interest in the establishment of an accurate and rapid

*Corresponding author: Email: pranitajaiswal@gmail.com.

methodology for the detection of contamination in food products, which should be cost-effective, simple, specific and capable of automation for online monitoring.

Keywords: IR Spectroscopy, food, HCA, PCA, PLSR, SIMCA

What Is Spectroscopy?

Electromagnetic spectrum is a diverse collection of radiant energy, from cosmic rays to X-rays to visible light to microwaves, each of which can be considered as a wave or particle traveling at the speed of light with different wavelengths and frequencies. Infrared waves form the part of electromagnetic spectrum between the visible and microwave regions. The IR region is divided into three regions: near infrared (12820-4000 cm–1), mid-IR (4000-400 cm–1), and far- IR (400- 33 cm–1). An increase in wave number corresponds to an increase in energy.

Vibration of chemical bonds between atoms in a molecule can be approximated by Hooke's Law. Two atoms and the connecting bond are treated as a simple harmonic oscillator composed of 2 masses (atoms) joined by a spring in which motion of each atom may be treated as an independent vibration with respect to a fixed centre of mass of the molecule. The vibration frequency is a function of the masses of the two atoms m1 and m2 and the strength of the bond k. There is a parabolic relationship between potential energy and interatomic distance.

In IR spectroscopy, an organic molecule is exposed to infrared radiation and when the frequency of the radiation matches that of the vibrating molecule, there occurs a net transfer of energy from the radiation to the molecule; this can be measured as a plot of energy versus wavelength, called a spectrum. The vibrational motion is quantized and it must follow the rules of quantum mechanics. The transitions that occur fit the following formula:

$$E = (n+1/2)\, h\nu$$

Where E is the photon energy, ν is the frequency of the vibration, n is the quantum number (0, 1, 2, 3, . . .), and h is a constant called Planck's constant.

Transitions are allowed only to the next energy level; therefore, molecules will absorb an amount of energy equal to 3/2 – 1/2 h ν or h ν. Occasionally, transitions of 2 h ν, 3 h ν, or higher are also observed, which are referred to as overtones in an IR spectrum. They are of lower intensity than the fundamental vibration bands. The intensity of overtone bands increases with the deviation of a particular bond system from harmonicity. The most anharmonic bonds are those involving the lightest atom, hydrogen, and a heavier atom such as carbon, nitrogen or oxygen. While each frequency of vibration of a complex molecule generally involves oscillation of all the atoms, the large amplitude motions are often limited to a particular functional group, e.g., -OH, -NH2. The characteristic frequency is similar for the same functional group where it appears in different molecules. This concept of characteristic group frequencies is the basis of analysis by infrared (IR). Each fundamental will be associated with its own potential energy curve and its own set of overtones. The number of bands a given molecule will produce can be calculated from the number of coordinates required to describe the total momentum of the molecule. A molecule consisting of n atoms has 3n degrees of freedom; therefore, 3n coordinates describe the location in space of a molecule of n atoms of which 3 describe the translational and 3 the vibrational motion (2 if linear) of the molecule. There are, therefore, 3n-6 (3n-5 if linear) possible modes of vibration or fundamental vibrations. There are two types of molecular vibrations, stretching and bending. Bending vibrations occur at lower frequencies than corresponding stretching vibrations.

In order to be IR active, a vibration must cause a change in the dipole moment of the molecule. In the NIR region, combination bands arise by interaction of two or more vibrations taking place simultaneously and the frequency of a combination band is simply the sum of multiples of the relevant fundamental frequencies. Thus, a triatomic molecule with 3 fundamental bands could have 3 first overtones, 3 second overtones and 6 combination bands. NIR spectra are much more complex than IR spectra, but overtones and combinations can still be assigned to specific group frequencies.

A molecule behaves as an anharmonic oscillator. The energy levels become more closely spaced with increase in interatomic distance in the anharmonic oscillator. The allowed transitions, hv, become smaller in energy. Therefore, overtones can be lower in energy than predicted by the harmonic oscillator theory. The equation below shows the relation among bond strength, atomic mass and wavenumber at which a molecule will absorb IR radiation.

$$\bar{v} = \frac{1}{2\pi c}\sqrt{\frac{f(m_1 + m_2)}{m_1 m_2}}$$

Where $\bar{v}$ is the vibrational frequency (cm–1)
m_1 and m_2 are the mass of atoms 1 and 2, respectively, in g
c is the velocity of light (cm/s)
f is the force constant of the bond (dyne/cm)

As the force constant increases, the vibrational frequency (wavenumber) also increases. The force constants for bonds are: single bond 5x105 dyne/cm, double bond 10x105 dyne/cm and triple bond 15x105 dyne/cm. As the mass of the atoms increases, the vibration frequency decreases. The actual range for C–H absorptions is 2850–3000 cm–1. The region of an IR spectrum where bond stretching vibrations are seen depends primarily on whether the bonds are single, double or triple, or bonds to hydrogen.

VIBRATIONAL SPECTROSCOPY TECHNIQUES

Vibrational spectroscopy is used to study a very wide range of sample types and can be carried out from a simple identification test to an in-depth, full spectrum, qualitative and quantitative analysis. Samples may be examined either in bulk or in microscopic amounts over a wide range of temperatures and physical states (e.g., gases, liquids, latexes, powders, films, fibers or as a surface or embedded layer). Vibrational spectroscopy has a very broad range of applications and provides solutions to a host of important and challenging analytical problems. The important vibrational spectroscopy techniques are mid-infrared (IR), near-IR, and Raman spectroscopy. Near-IR spectroscopy measures the broad overtone and combination bands of some of the fundamental vibrations. Both mid-IR and Raman spectroscopy provide characteristic fundamental vibrations that are employed for the elucidation of molecular structure. All three techniques have various advantages and disadvantages with respect to instrumentation, sample handling, and applications.

Raman and mid-IR spectroscopy are complementary techniques and usually both are required to completely measure the vibrational modes of a molecule. Although some vibrations may be active in both Raman and IR, these two forms of spectroscopy arise from different processes and different selection rules. In general, Raman spectroscopy is best at symmetric vibrations of non-polar groups while IR spectroscopy is best at the asymmetric vibrations of polar groups. Infrared and Raman spectroscopy involve the study of the interaction of radiation with molecular vibrations, but differs in the manner in which photon energy is transferred to the molecule by changing its vibrational state. IR spectroscopy measures transitions between molecular vibrational energy levels as a result of the absorption of mid-IR radiation. This interaction between light and matter is a resonance condition involving the electric dipole-mediated transition between vibrational energy levels. Raman spectroscopy is a two-photon inelastic light-scattering event. Here, the incident photon is of much greater energy than the vibrational quantum energy, and loses part of its energy to the molecular vibration with the remaining energy scattered as a photon with reduced frequency. In the case of Raman spectroscopy, the interaction between light and matter is an off-resonance condition involving the Raman polarizability of the molecule. The IR and Raman vibrational bands are characterized by their frequency (energy), intensity (polar character or polarizability), and band shape (environment of bonds). Since the vibrational energy levels are unique to each molecule, the IR and Raman spectrum provide a "fingerprint" of a particular molecule. The frequencies of these molecular vibrations depend on the masses of the atoms, their geometric arrangement, and the strength of their chemical bonds. The spectra provide information on molecular structure, dynamics and environment.

CHEMOMETRICS

Spectral Pre-Processing

As most of bacterial spectra resemble considerably, the IR spectra of bacteria is generally pre-processed using various manipulations to facilitate spectral interpretation and analysis (Smith 1996). Spectral pre-processing is generally performed with the software that comes with a spectrometer or other software like the Unscrambler. Some of the pre-processing methods used for bacterial spectra are described below:

Baseline correction and smoothing

Baseline correction

The dissimilarities in spectra due to shifts in baseline are eliminated using baseline correction. Smoothing reduces the high frequency noise in spectral signal.

Derivatives

The first and second derivative transformations have been used in analytical spectroscopy to reduce replicate variability, remove baseline sloping effects, resolve overlapping absorption bands and amplify spectral variations. First derivative transformation is generally used for the taxonomic classification of bacteria (Rebuffo-Scheer et al. 2007). Second derivative spectra increase the number of discriminative features associated with bacterial spectra and are commonly used for bacterial classification (Rodrigues-Saona et al. 2001; Alexandrakis et al. 2008; Al-Holi et al. 2006), and detecting sub-lethal injury of bacteria due to heat, chlorine, radicals and sonication (Lorin-Latxague & Melin 2005; Al-Qadiri et al. 2008b).

Normalization

Normalization attempts to correct differences arising from path length variations, scattering effects, source or detector variations, or other general instrumental sensitivity effects. Normalization also helps give all samples an equal impact on the model. Typically, normalization should be performed before any centering or scaling or other column-wise pre-processing steps and after baseline or offset removal. The presence of baseline or offsets can impede correction of multiplicative effects. The effect of normalization prior to background removal will, in these cases, not improve model results and may deteriorate model performance (Alexandrakis et al. 2009). The spectra are normalized to the most intense band or at the same integrated intensity in a given spectral region. Usually, the amide I band is used as an internal standard for normalization (Yu & Irudayaraj 2005).

Statistical Analysis/Chemometrics

Chemometrics refers to a combination of multivariate statistical and mathematical approaches used to extract useful information from complex spectra. Different chemometric methods used are Principal Component Analysis (PCA), Hierarchical Cluster Analysis (HCA) PCA, Discriminant Analysis (DA), Canonical Variate Analysis (CVA) and Artificial Neural Network (ANN) analysis. In the case of PCA and HCA, prior knowledge about the type of bacteria studied is not essential while for DA, CVA and ANN, information about the sample identity is required. With later methods, a model can be trained so that it can predict the identity of unknown samples.

Principal Component Analysis (PCA)

PCA identifies patterns or relationships in the dataset. This method reduces the multidimensionality of spectral data by projecting a large number of original variables into few latent variables known as principal components. The first principal component (PC) expresses the largest proportion of variation, followed by the second PC that conveys the second most important factor of the remaining analysis (Rodriguez-Saona et al. 2004, Alexandrakis et al. 2008, Alexandrakis et al. 2009), and so forth. Score plots can be used to interpret the similarities and differences between bacteria. The closer the samples are within a score plot, the more similar they are with respect to the principal component score evaluated (Al-Holy et al. 2006).

Hierarchical Cluster Analysis (HCA)

On the basis of distances between spectra and the linkage algorithm used, HCA identifies similarities between the spectra of microorganisms. The frequently used linkage distances metric are the Pearson product moment correlation coefficient and the Euclidian distance. Ward's algorithm and the average linkage algorithm (also called UPGMA, the Unweighted Pair Group Method with Arithmetic Mean) are the commonly used cluster analysis algorithms for microbial identification (Beveridge & Graham, 1991). Results of HCA are presented in the form of a dendrogram. The left vertical axis of a dendrogram depicts the increasing variance or heterogeneity. The magnitude of this heterogeneity depends on the number of spectra in a cluster and the similarities between them (Davis & Mauer 2010). HCA can also be used to identify an unknown bacterium by calculating the spectral distance between it and a known bacterium (Kummerle et al. 2008). Schuester et al. (1999) used HCA to characterize the physiological status of a microbial culture in a biotechnological process.

Canonical Variate Analysis (CVA) and Discriminant Analysis

CVA or canonical discriminant analysis is a widely used method for analyzing group structure in multivariate data having more than two groups. When there are only two groups to be separated, the technique is known as discriminant function analysis (DFA). CVA groups data by estimating the canonical variates, which are linear combinations of the original variable. Canonical variate analysis gives a graphical representation of the similarities and differences between spectra (Yang & Irudayaraj 2003). Discriminant analysis is another multivariate statistical technique for the classification of objects into clusters by determining the similarity of a set of values from an unknown sample to a set of values measured from a set of known samples.

Soft Independent Modelling by Class Analogy (SIMCA) and Artificial Neural Network (ANN)

SIMCA requires a training dataset consisting of samples with a set of attributes and their class membership. The term soft refers to the fact that the classifier can identify samples as belonging to multiple classes and not necessarily producing a classification of samples into non-overlapping classes. Before SIMCA modeling, a PCA model is made for each class. Unknown samples are then compared to the class models. SIMCA models were used to classify bacteria (Al-Qadiri et al. 2006), discriminate spectra of bacteria at different growth phases (Al-Qadiri et al. 2008a) and also to differentiate between viable and sub-lethally injured bacteria (Al-Qadiri et al. 2008b). ANN, like SIMCA, also requires a training dataset and a validation dataset. ANN-based analysis was reported for the classification of Listeria to the species level and L. monocytogenes to serogroup and serovar levels (Rebuffo-Scheer 2007). For the ANN-based classification, HCA was used as a validation method (Davis and Mauer 2010).

Partial Least Squares Regression (PLSR)

PLSR is a recently developed generalization of multiple linear regression method (MLR). PLSR, unlike MLR, can analyze data with strongly collinear, correlated, noisy, and numerous X-variables and also simultaneously model several response variables, Y. In PLSR, PLS components are extracted from a data matrix X in order to explain as much of the variation in a reference matrix Y as possible while at the same time accounting for the variation in the data matrix X (Ellis et al. 2002; Brandily et al. 2011). When PLSR is used for discrimination, it is called discriminant PLSR. In order to build a PLSR model, a calibration dataset (known sample), a validation dataset, and a test sample are needed. Discriminant PLSR models were used to classify and identify different species of pathogenic bacteria (Alexandrakis et al. 2008; 2009).

APPLICATIONS IN MICROBIOLOGY

Conventional microbial identification techniques are elaborate and several tests are required to identify bacteria, even at the genus level, which may each take several days. Although kits are more efficient, they are expensive and are available only for selected bacteria. Automated procedures like gas chromatography and protein electrophoresis are available, but these are time consuming and require elaborate sample preparation and trained manpower. Therefore, food industry is continuously in search of rapid, reliable, specific and inexpensive methods for enumerating and identifying microorganisms in food products and processing equipments. Spectroscopic methods are routinely used within the food industry as part of their quality control protocols. These methods are rapid, cost-effective, reliable and non-invasive. The expansion of these tools to include screening for possible microbial contamination is becoming more feasible due to the research advances during the last two decades.

DETECTION AND IDENTIFICATION

Bacterial identification and differentiation by mid infrared spectroscopy (2500-5000nm) has been investigated since the middle of the last century, but shortcomings in available instrumentation led to discontinuation of these studies for several decades. However, Naumann et al. (1991) reported the use of Fourier-transform infrared (FTIR) spectroscopy to chemically discriminate intact microbial cells producing complex biochemical fingerprints that are reproducible and distinct for different bacteria more than a decade ago. The level of resolution of Fourier transform infrared (FTIR) absorbance spectroscopy (in the mid-IR range, usually defined as 4000-400 cm-1 or 2500-25000 nm) has been found to be sufficient to distinguish intact microbial cells at the strain level (Helm et al. 1991). The quantitative analysis of the cellular contents for bacterial cells was reported to be possible by the application of FT-IR in combination with multivariate statistical techniques (Helm et al. 1991;Kansiz et al. 1999). The resulting spectra had been shown to reflect the "total" biochemical composition of the microbial cell (Kansiz et al. 1999). FT-IR spectroscopy is based on vibrations of functional groups and highly polar bonds. These spectral features of many important biochemical constituents create a "fingerprint" for a bacterial strain. For microbial samples, these spectral elements are most commonly DNA/RNA, proteins, and membrane and cell wall amine- and fatty acid-containing components (Goodacre et al. 1996). More recently, the potential for FTIR to reproducibly discriminate between closely related bacterial species (Rodriguez- Saona et al. 2001) has been shown. Lin et al. (2004) demonstrated the application of SW-NIR spectrometry together with chemometrics in accurate quantification of the bacterial loads in chicken muscle. They reported that this could be carried out using the PLS-based prediction method. Lin et al. (2006) further demonstrated use of Principal Component Analysis (PCA) and Partial Least Squares (PLS) chemometric models to predict the onset and degree of spoilage. Al-Qadiri et al. (2006) reported that identification of primary bacterial contaminants P. aeruginosa and E. coli. in drinking water can be done with FTIR spectroscopy and multivariate analysis (PCA and SIMCA). They could further identify different segments of wavelength, which showed variations in bands depicting amides, phosphodiester backbone of nucleic acids, polysaccharide and proteins for the bacterial contaminants. They reported that the greatest spectral variation between these microbes was observed between 1800 and 900 cm-1. Similarly, detection of bacteria causing mastitis infection in rabbit milk was reported using NIRS in combination with Soft Independent Modeling of Class Analogy (SIMCA) (Daskalov et al. 2011). Ellis et al. (2002) has shown that the microbial count and spoilage in chicken can be monitored by FTIR, suggesting that proteolysis is the main indicator for the onset of spoilage when microorganisms have reached 107cfug-1.

Discrimination of Viable, Injured, and Dead Bacteria

Many microbiological applications are based on accurate determination of live, dead, and total bacteria in a sample of interest, as injured bacteria can cause a significant health threat if they revive during food distribution and storage and it is important to examine the efficacy of various intervention treatments used in food processing. IR spectroscopy offers a simple and cost-effective solution to the problem of differentiating live and dead cells, as it is based on the changes in biochemical composition of cells. Lin et al. (2004) analyzed intact as well as sonicated cells of L. monocytogenes ATCC 19114 using FTIR spectroscopy, observing them in the spectral range of 1500 and 800 cm-1. The FTIR vibrational overtone and combination bands from mid-IR active components of intact and injured bacterial cells produced distinctive "fingerprints." They could obtain clear segregations with distinct sample clusters between all intact strains and the injured L. monocytogenes ATCC 19114 using PCA on spectral data in the fingerprint region. This indicated that FT-IR can discriminate between different strains in the genus Listeria and between intact and injured cells of this genus. The application of statistical transformations to the spectral data of sub-lethally heat-injured microorganisms: Salmonella enterica serotype Typhimurium ATCC 14028, a Gram-negative bacterium, and Listeria monocytogenes ATCC 19113, a Gram-positive bacterium, showed differences, indicating the mechanism of cell injury to be different for Gram-negative and Gram-positive microbes (Al-Qadiri et al. 2008a). Further, the extent of injury done to the microbial cell in a range of heat treatments could be predicted with the use of FTIR spectroscopy together with chemometrics (SIMCA) (Al-Qadiri et al. 2008b). Irudayaraj et al. (2010) demonstrated that a good PLS prediction model could be developed for establishing the relationship between different ratios of live and dead E. coli O157:H7 and FT-IR spectra. Further, they used canonical variate analysis (CVA) for separating the spectra of 100% dead cells from ground beef from the spectra of 0.5% live cells in the presence of 99.5% dead cells of E. coli O157:H7. This indicated the utility of FTIR methods for rapid detection and differentiation of pathogens in complex foods.

Analysis of Structural Components of Bacteria

Infrared spectroscopy has been used to study membrane composition of bacteria since early decades of the twentieth century. Whittaker et al. (2003) demonstrated the use of spectroscopy while measuring FAME mixtures for rapid (2 min) identification of bacteria. Stuart (2004) differentiated intact microbial cells based on the biochemical heterogeneity of outer membrane cell components such as lipoproteins, lipopolysaccharides (LPS) and phospholipids using FTIR spectroscopy. Spectra of crude LPS extracts from E. coli and S. enterica were used for serotype level differentiation with 95-100% accuracy (Kim et al. 2005; 2006a). Kim et al. (2006b) also differentiated serotypes of Salmonella enterica based on outer membrane proteins using FTIR spectroscopy and chemometrics.

Biological Process Monitoring

Near-infrared (NIR) spectroscopy has been used to provide on-line information on substrate, biomass, product, and metabolite concentrations in fermentation processes, which could be useful for improved monitoring or control (Cervera et al. 2009). However, several factors can negatively influence the quality of chemometric models built for interpretation of the spectra, thus impairing the analyte concentration predictions. Tosi et al. (2003) demonstrated the use of partial least squares regression (PLSR) calibration models of the second derivative NIR spectra in the range of 700-1800 nm for prediction of biomass. Recently, Scholz et al. (2011) demonstrated the scope of FTIR spectroscopy in drug inactivation studies and plasmid production processes.

References

Alexandrakis D, Downey G, Scannell AGM. Detection and identification of bacteria in an isolated system with near infrared spectroscopy and multivariate analysis. *J Agri Food Chem* 2008; 56(10): 3431-37.

Alexandrakis D, Downey G, Scannell AGM. Rapid non-destructive detection of spoilage of intact chicken breast muscle using near-infrared and fourier transform using mid near-infrared spectroscopy and multivariate statistics. *Food Bioprocess Technol* 2009; DOI 10.1007/s11947-009-0298-4.

Al-Holi MA, Lin M, AL-Qadiri H, Cavinato AG, Rasco BA. Classifications of foodborne pathogens by Fourier transform infrared spectroscopy and pattern recognition techniques. *J Rapid Methods Autom Microbiol* 2006; 14:189-200.

Al-Holy M, Lin M, Cavinato AG, Rasco BA. The use of Fourier transform infrared spectroscopy to differentiate *Escherichia coli* O157:H7 from other bacteria inoculated into apple juice. *Food Microbiol* 2006; 23:162-168.

Al-Qadiri H, Lin M, Cavinato AG & Rasco BA. Fourier transform infrared spectroscopy, detection and identification of Escherichia coli O157:H7 and Alicyclobacillus strains in apple juice. *Int J Food Microbiol* 2006; 111(1): 73–80.

Al-Qadiri HM, Al-Alami N, Al-Holy MA, Rasco BA. Using Fourier transform infrared (FT-IR) absorbance spectroscopy and multivariate analysis to study the effect of chlorine-induced bacterial injury in water. *J Agri Food Chem* 2008a; 56:8992-97.

Al-Qadiri HM, Lin M, Al-Holy M, Cavinato AG & Rasco BA. Detection of sublethal thermal injury in Salmonella enterica serotype Typhimurium and Listeria monocytogenes using Fourier transform infrared (FT-IR) spectroscopy (4000 to 600 cm−1). *J Food Sci* 2008b; 73(2): M54–M61.

Beveridge TJ, Graham LL. Surface layers of bacteria. *Microbiol. Rev.* 1991; 55:684-705.

Brandily ML, Monbet V, Bureau B et al. Identification of foodborne pathogens within food matrices by IR spectroscopy. *Sens and Actu B* 2011; 160: 202-06.

Cervera AE, Petersen N, Lantz AE, Larsen A, Gernaey KV. Application of near-infrared spectroscopy for monitoring and control of cell culture and fermentation. *Biotechnol Prog* 2009; 25(6):1561-81.

Daskalov H, Atanassova S, Stoyanchev T & Santo R. "Application of near infrared spectroscopy for rapid non-invasive detection of *Listeria monocytogenes*, *Escherichia coli* and *Staphylococcus aureus* growth in foods". *Bulgarian J Vet Med* 2011; 14(3): 150–157.

Davis R, Irudayaraj J. Reuhs BL, Mauer LJ. Detection of E. coli O157:H7 from ground beef using Fourier transform infrared (FT-IR) spectroscopy and chemometrics. *J Food Sci*. 2010; 75(6):M340-6.

Davis R, Mauer LJ (2010). Fourier transform infrared (FT-IR) spectroscopy: a rapid tool for detection and analysis of foodborne pathogenic bacteria. In: Méndez-Vilas A (ed) *Current research, technology and education topics in applied microbiology and microbial biotechnology*. Formatex Research Center, Spain. pp 1582–1594.

Ellis DI, Broadhurst D, Kell DB, Rowland JJ & Goodacre R. Rapid and quantitative detection of the microbial spoilage of meat by Fourier transform infrared spectroscopy and machine learning. *Appl Environ Microbiol* 2002; 68(6): 2822–28.

Goodacre R, Timmins EM, Rooney PJ, Rowland JJ, Kell DB. Rapid identification of *Streptococcus* and *Enterococcus* species using diffuse reflectance-absorbance Fourier transform infrared spectroscopy and artificial neural networks. *FEMS Microbial Lett* 1996; *140:* 233-239.

Helm D, Labischinski H Schallehn G, Naumann D. Classification and identification of bacteria by Fourier transform infrared spectroscopy. *J Gen Microbiol* 1991; *13:* 69-79.

Kansiz M, Heraud P, Wood B, Burden F, Beardall J, McNaughton D. Fourier transform infrared microspectroscopy and chemometrics as a tool for the discrimination of cyanobacterial strains. *Phytochemistry* 1999; *52:* 407-417.

Kim S, Burgula Y, Ojanen-Reuhs T, Cousin MA, Reuhs BL, Mauer LJ. Differentiation of crude lipopolysaccharides from *Escherichia coli* strains using Fourier transform infrared spectroscopy and chemometrics. *J Food Sci* 2006a; 71:M57-M61.

Kim S, Kim H, Reuhs BL, Mauer LJ. Differentiations of outer membrane proteins from Salmonella enterica serotypes using Fourier transform infrared spectroscopy and chemometrics. *Lett Appl Microbiol* 2006b; 42:229-234.

Kim S, Reuhs BL, Mauer LJ. Use of Fourier transforms infrared spectra of crude bacterial lipopolysaccharides and chemometrics for differentiation of Salmonella enterica serotypes. *J Appl Microbiol* 2005; 99:411-417.

Kummerle M, Scherer S. & Seiler H.Rapid and reliable identification of food-borne yeasts by fourier-transform infrared spectroscopy. *Appl Environ Microbiol* 1998; 64(6): 2207-14.

Lin M, Al-Holy M, Al-Qadiri H, Kang D, Cavinato AG, Huang Y, et al. Discrimination of intact and injured Listeria monocytogenes by Fourier transform infrared spectroscopy and principal component analysis. *J Agri Food Chem* 2004; 52(19): 5769–72.

Lin, M., M. Mousavi, M. Al-Holy, A. Cavinato, and B. A. Rasco. 2006. Rapid near infrared spectroscopic method for the detection of spoilage in rainbow trout (Oncorhynchus mykiss) fillet. *J. Food Sci.* 71:18-23.

Lorin-Latxague C, Melin AM. Radical induced damage of *Micrococcus luteus* bacteria monitored using FT-IR spectroscopy. *Spectroscopy*. 2005; 19:17-26.

Naumann D, Helm D & Labischinski H. Microbiological characterizations by FT-IR spectroscopy. *Nature* 1991; 351(6321): 81–82.

Rebuffo-Scheer CA, Schmitt J, Scherer S. Differentiation of *Listeria monocytogenes* serovars by using artificial neural network analysis of Fourier-transformed infrared spectra. *Appl Environ Microbiol* 2007; 73:1036-1040.

Rodriguez-Saona LE, Khambaty F, Fry F, Dubois J, Calvey EM. Detection and identification of bacteria in a juice matrix with Fourier transform-near infrared spectroscopy and multivariate analysis. *J Food Prot* 2004; 67: 2555-59.

Rodriguez-Saona LE, Khambaty FM, Fry FA, Calvy EM. Rapid detection and identification of bacterial strains by Fourier transform near-infrared spectroscopy. *J Agric Food Chem* 2001; 49: 574–79.

Scholz T, Lopesy V V, Caladoz C RC. Monitoring bacterial processes by Fourier transform infrared spectroscopy: Helicobacter pylori drug inactivation and plasmid bioproduction in recombinant Escherichia coli cultures In: *1st Portuguese Meeting in Bioengineering*, February 2011 Portuguese chapter of IEEE EMBS Instituto Superior T´ecnico, Technical University of Lisbon.

Schuster KC, Mertens F, Gapes JR. FTIR spectroscopy applied to bacterial cells as a novel method for monitoring complex biotechnological processes. *Vib Spec* 1999; 19:467-477.

Smith B. *Fundamentals of Fourier Transform Infrared Spectroscopy.* CRC Press, Washington, D.C. 1996.

Stuart B. Infrared spectroscopy. In: *Fundamentals and Applications*. New York: Wiley.

Tosi S, Rossi M, Tamburini E, Vaccari G, Amaretti A, Matteuzzi D. Assessment of in-line near-infrared spectroscopy for continuous monitoring of fermentation processes. *Biotechnol Prog*. 2003; 19(6):1816-21.

Whittaker P, Mossoba MM, Al-Khaldi S, Fry FS, Dunkel VC, Tall BD, Yurawecz MP. Identification of foodborne bacteria by infrared spectroscopy using cellular fatty acid methyl esters. *J Microbiol Methods* 2003; 55:709– 16.

Yang H & Irudayaraj J. Rapid detection of food borne microorganisms on food surface using Fourier transform Raman spectroscopy. *J Mol Struc* 2003; 646: 35-43

Yu C, Irudayaraj J. Spectroscopic characterization of microorganisms by Fourier transform infrared microspectroscopy. *Biopolymers* 2005; 77:368-77.

In: Recent Advances in Microbiology. Volume 3 ISBN: 978-1-53614-057-6
Editors: N.C. Gautam, S.P. Tiwari and R. Sharma

Chapter 9

BRAIN-INFECTING MICROORGANISMS

Vaibhav Mishra[1,2,*], Akanksha Srivastava[1], Haushila Prasad Pandey[2] and Gautam Palit[1]

[1]Division of Pharmacology, CSIR-Central Drug Research Institute, Lucknow, UP, India

[2]Department of Biochemistry, Banaras Hindu University, Varansi, UP, India

INTRODUCTION

Several microorganisms are capable of infecting the brain and its associated tissues. Bacteria and viruses are the most common microorganisms that infect brain. Fungi, parasites and others can also infect the brain, though rarely. Germs infect the brain, causing an inflammation of the invaded area and manifesting the disease. The nomenclature of the brain infections vary, depending upon the exact part of the brain that has been targeted by the microorganism:

Meningitis- Meningitis is the inflammation of the meninges, which is the three-layered membrane enclosing the brain and spinal cord; the fluid it is bathed in is called the cerebrospinal fluid (CSF). (Llorens & McCracken, 2003).

- *Encephalitis*- It is an inflammation of the brain itself.
- *Myelitis*- Spinal cord inflammation is known as Myelitis.
- *Abscess*- Accumulation of infectious material and offending microorganisms within the CNS.

* Corresponding author E-Mail: vaibhavpharma06@gmail.com.

Meningitis may be either of viral or bacterial origin:

1. *Viral meningitis* requires no specific treatment, and goes away completely without complications.
2. *Bacterial meningitis* is a very serious disease and may result in speech faults, hearing loss, seizures, loss of extremity function or deduction, permanent brain damage, and even death. Bacterial meningitis is of great concern worldwide and may be a serious threat to global health.

Causes of Bacterial Meningitis

Three bacterial species are the most common causes of meningitis in all age groups, except newborns:

- *Streptococcus pneumonia* (causing pneumococcal meningitis)
- *Neisseria meningitidis* (causing meningococcal meningitis)
- *Haemophilus influenza* type b (Hib)

How Organisms Are Transmitted

The bacteria may be spread through the exchange of respiratory and throat secretions, from coughing, sneezing, or kissing. The only exception is meningococcal meningitis (Ginsberg, 2004).

Bacterial Meningitis: The Most Susceptible Groups

About two-third cases of bacterial meningitis are infants and small children. It is more common in males than in females and is more likely seen in late winter and early spring. Patients include day-care workers, military recruits, jail cellmates, and anyone directly exposed to discharges from the mouth or nose of an infected person (Hartman & Kaplan, 2004). The other groups at risk include people with weakened immune systems, diabetics, chronic alcoholics, IV drug abusers, and anyone older than 60 years of age.

Encephalitis- Encephalitis or Japanese encephalitis is caused by a virus from the family Flaviviride. Pigs and wild birds are reservoirs of the virus; transmission to humans may cause severe symptoms. The most important vectors of this disease are the mosquitoes *Culex tritaeniorhynchus* and *Culex vishnui*. This disease is most prevalent in south-east Asia.

Symptoms and Signs

In the early onset of the disease, Japanese encephalitis shows non-specific symptoms such as fever, headache and malaise in humans. Signs that develop during the acute encephalitic stage include neck rigidity, cachexia, hemiparesis, convulsions and high body temperature between 38 and 41 degrees Celsius. Mental retardation developed from this disease usually leads to coma (Gupta et al., 2010). Mortality of this disease varies, but is usually much elevated in children. Transplacental spread has been noted. Lifelong neurological defects such as deafness and emotional lability may occur in patients who suffered central nervous system involvement. In some cases, symptoms also include nausea, headache, vomiting, fever and occasionally, swelling of the testicles (Gupta & Rao, 2011). Increased microglial activation following Japanese encephalitis infection has been found to influence the outcome of viral pathogenesis. Microglias are the resident immune cells of the central nervous system and have a critical role to play in host defence against the invading microorganisms. Activated microglia secrete cytokines, such as interleukin-1 (IL-1) and tumor necrosis factor alpha (TNF-α), which can cause toxic effects in the brain. Additionally, other soluble factors such as neurotoxins, excitatory neurotransmitters, prostaglandin, reactive oxygen, and nitrogen species are secreted by the activated microglia.

Treatment

Mannitol controls the increased intracranial pressure (Jmor et al., 2008). Japanese encephalitis is not transmitted from one to another; therefore, the patients need not be isolated. A breakthrough in the field of Japanese encephalitis therapeutics is the identification of macrophage receptor for Japanese encephalitis. A recent report demonstrates the involvement of monocytes and macrophage receptors in severe inflammatory response in Japanese encephalitis infection of the brain. This study provides a hypothesis of neuro-inflammation and in fact, a new lead in the development of appropriate therapeutics against Japanese encephalitis (Rozenberg et al., 2011) [7].

Myelitis- It is the infection or the inflammation of white matter or gray matter of spinal cord. During an inflammatory response from the spinal cord, the myelination and axons may be damaged, causing symptoms such as paralysis and sensory loss.

Types of Myelitis

Myelitis lesions usually occur in a narrow region, but can spread and affect many areas.

- *Poliomyelitis:* disease caused by viral infection in the gray matter with symptoms of muscle paralysis or weakness (Kelly, 2006).
- *Leukomyelitis:* lesions in the white matter.
- *Transverse Myelitis:* caused by axonal demyelination encompassing both sides of the spinal cord.
- *Meningococcal Myelitis or Meningomyelitis:* lesions occurring in the region of meninges and the spinal cord.

Symptoms

Muscle weakness or paralysis starts with the legs and then reaches to the arms. General symptoms include paresthesia, which is a sensation of tickling, tingling, burning, pricking, or numbness of a person's skin with no apparent long-term physical effects. Adult patients often report pain in the back, legs and arms or stomach (Homas & Thomas, 1997). Patients also present increased urinary urgency, bowel or bladder dysfunctions such as bladder incontinence, difficulty or inability to void, and incomplete evacuation of bowel or constipation. Others also report fever, respiratory problems and intractable vomiting (Wasay et al., 2006).

Treatment

Intravenous Steroids (methyl-prednisolone) is the only treatment for myelitis. If intravenous steroid is not effective, then one should treat with plasma exchange (PLEX). It has also been shown to be effective with other autoimmune or inflammatory central nervous system disorders (Scott et al., 2011).

Immuno-suppressants/Immuno-modulatory Agents: Myelitis with no definite cause seldom recurs, but in other cases, myelitis may be a manifestation of other diseases that are mentioned above. In these cases, ongoing treatment with medications that modulate or suppress the immune system may be necessary. Or, sometimes there is no specific treatment. Either way, aggressive rehabilitation and long-term symptom management are an integral part of the healthcare plan.

Abscess: Brain abscess or cerebral abscess is caused by inflammation and collection of infected material, coming from ear infection, dental abscess, infection of paranasal sinuses, infection of the mastoid air cells of the temporal bone, epidural abscess or even from remote organs such as the lung, heart or kidney. infectious sources, within the brain tissue. The infection may also be introduced through a skull fracture following a head trauma or surgical procedures. Brain abscess is usually associated with congenital heart disease in young children. It may occur at any age, but is most frequent in the third decade of life. Deadly brain abscesses due to infection traced to tongue piercings have also been observed.

Diagnosis

The diagnosis is established by a computed tomography. At the initial phase of the inflammation (which is referred to as cerebritis), the immature lesion does not have a capsule and it may be difficult to distinguish it from other space-occupying lesions or infarcts of the brain. Lumbar puncture procedure, which is performed in many infectious disorders of the central nervous system, is challenging in this condition. Ring enhancement may also be observed in cerebral hemorrhages and some brain tumors. However, in the presence of the rapidly progressive course with fever, focal neurologic findings (hemiparesis, aphasia etc.) and signs of increased intracranial pressure, the most likely diagnosis should be brain abscess (Brook, 1995).

Treatment

The treatment includes the lowering of the increased intracranial pressure and starting intravenous antibiotics, with simultaneous efforts in identifying the causative organism, mainly through blood culture studies.

Surgical drainage and treatment of the primary lesion is crucial, as is the removal of any foreign material (bone, dirt, bullets, and so forth).

Other Common Brain Infections

Toxoplasmosis is caused by the parasite *Toxoplasma gondii* (cat is a host for this organism). Infection is spread by several means; for example, from an infected mother to an unborn baby, by eating unwashed vegetables or uncooked meat, or by coming in direct contact with cat feces, and among patients with a weakened immune system such as HIV positive patients.

Cerebral Cysticercosis is caused by the pork tapeworm. When people eat food contaminated by feces or eat the larvae, they mostly suffer from this disease. The most common symptoms are seizures. Few medications may stop the progression of the disease. However, once the cerebral form is acquired, the treatment is usually given to relieve the symptoms.

Trichinosis is caused by the roundworm *Trichinella spiralis*. It is spread by eating larvae in raw or undercooked pork. An infected person may show symptoms similar to encephalitis with confusion and delirium. Coma, seizures, paralysis, and other signs of neurological loss are found in more severe forms.

German Measles or Rubella is caused by the rubella virus. This disease may affect the unborn fetus that may be born with a variety of defects including deafness, mental retardation, and heart problems. At birth, the infant has meningitis-like illness and is usually lethargic and inactive. Proper immunization of the mother with a series of vaccinations given throughout adolescence and early adulthood would prevent a woman from getting rubella, which is especially important during pregnancy.

Mumps and Measles are both caused by viruses. Young children are most commonly affected. Transmission occurs through respiratory routes. Complications may include viral meningitis or encephalitis in varying degrees of severity. The most common complications of mumps and measles are deafness and seizures, respectively. Prevention may be attained through adequate childhood immunization.

Rabies is another viral infection of the CNS. It is transmitted to humans by the bite of an infected mammal or in rare instances, by an inhalation of airborne viral particles in bat-infested caves or by laboratory workers. Worldwide, the disease is commonly caused by the bite of the rabid dog but may also be transmitted by cats, raccoons, skunks, foxes, wolves, and many other domestic and wild animals. Despite popular belief, no transmission has occurred from the bite of the mouse, rat, or rabbit. The virus causes a severe form of flu-like symptoms, very high (up to 107 ^{0}F) fever, extreme restlessness, hypersensitivity to touch, general convulsions, total body paralysis, bizarre hallucinations, excessive flow of saliva, absolute refusal to drink any fluids, gradual paralysis, coma, and almost invariably, death. There is no specific antiviral therapy available, but preventive immunization is very effective and widely available.

AIDS (and *HIV)* encephalitis is caused by the human immunodeficiency virus (HIV). HIV can directly infect the central nervous system, causing a range of neurologic conditions. The most common is so-called AIDS dementia. It is characterized by the slow onset of behavioral, intellectual, and motor impairment. Early symptoms include confusion, loss of libido, social withdrawal, decreased concentration, poor balance, and weaknesses. In the later stage, severe dementia, inability to control urine flow, and occasionally, an inability to speak and walk have been observed. The treatment includes standard antiviral drugs.

Brain Infection Treatment

Self-Care at Home

- Provide cooling measures and give temperature-lowering medications to reduce fever.

- If the person is vomiting, put him or her on their side to prevent them from inhaling and choking on the vomit.
- Avoid any strenuous activities, and keep the person on strict bed-rest.

Prevention of Brain Infection by Vaccines

- Hib vaccines are very safe and highly effective. It is a part of standard immunization for infants and children. The introduction of Hib vaccine as part of routine paediatric immunization has significantly reduced the occurrence of serious Hib disease. Newborns are usually infected with coli form such as *Escherichia coli* or *Listeria.*
- A vaccine against pneumococcal meningitis can also prevent other forms of infection.
- A vaccine against meningococcal meningitis is available.

References

Brook, I. Brain abscess in children: microbiology and management. *J. Child Neurol.*1995; 10 (4): 283–8.

Ginsberg, L. Difficult and recurrent meningitis. *Journal of Neurology, Neurosurgery, and Psychiatry*, 2004;75 Suppl 1 90001.

Gupta, N. and Rao, PVL. "Transcriptomic profile of host response in Japanese encephalitis virus infection." *Virology Journal*, 2011; 8 (92).

Gupta, N., Lomash, V. and Rao, PVL. Expression profile of Japanese encephalitis virus induced neuroinflammation and its implication in disease severity. *Journal of Clinical Virology* 2010; 49 (1): 04–10.

Hartman, B.J., Kaplan, S.L. Practice guidelines for the management of bacterial meningitis. *Clinical Infectious Diseases* 2004; 39 (9): 1267–84.

Homas, M., and Thomas, J. Acute transverse myelitis. *The Journal of The Louisiana State Medical Society*: Official Organ Of The Louisiana State Medical Society, 1997; 149(2): 75-77.

Jmor, F., Emsley, H.C., Fischer, M. The Incidence of acute encephalitis syndrome in western industrialised and tropical countries. *Virology Journal, 2008;* 5 (134): 134.

Kelly, H. Evidence for a causal association between oral polio vaccine and transverse myelitis: A case history and review of the Literature. *Journal of Paediatrics and Child Health,* 2006; 42(4): 155-159.

Llorens, S.X., McCracken, G.H. Bacterial meningitis in children. *Lancet,* 2003; 361 (9375): 2139-48.

Rozenberg, F., Deback, C., Agut, H. Herpes simplex encephalitis: from virus to therapy. *Infectious Disorders Drug Targets,* 2011; 11 (3): 235–250.

Scott, T., Frohman, E., De Seze, J., Gronseth, G., and Weinshenker, B. Evidence-based guideline: clinical evaluation and treatment of transverse myelitis: report of the Therapeutics and Technology Assessment Subcommittee of the American Academy of Neurology. *Neurology*, 2011; 77(24): 2128-2134.

Wasay, M., Arif, H., Khealani, B. and Ahsan, H. Neuroimaging of tuberculous myelitis: analysis of ten cases and review of literature. *Journal of Neuroimaging: Official Journal of The American Society of Neuroimaging*, 2006; 16(3): 197-205.

In: Recent Advances in Microbiology. Volume 3 ISBN: 978-1-53614-057-6
Editors: N.C. Gautam, S.P. Tiwari and R. Sharma

Chapter 10

NANOTECHNOLOGY IN THE DETECTION OF PATHOGENIC MICROORGANISMS

Pradeep Kumar** and Vandana Rai
Department of Biotechnology,
VBS Purvanchal University, Jaunpur, India

INTRODUCTION

One of the major health care problems of human beings is infectious diseases. The majority of these diseases are caused by contaminated food, water and air (Taubes, 2008; Kaittanis et al., 2010). They are responsible for significant human pathogenesis, morbidity and mortality around the world (Taubes, 2008; Hauck et al., 2010). Infectious diseases initiated in a particular region of the world can rapidly spread around the globe. The most effective strategy to prevent the spread of an infectious disease is rapid and accurate identification of the pathogen. Early diagnosis helps in control and eradication of the infectious disease. It also gives sufficient time to government and clinicians to take necessary preventive measures for disease-zoning (Tallury et al. 2010). On the other hand, misdiagnosis and delay in diagnosis of potentially dangerous pathogens is a limitation in early control of the disease. During the last century, significant advancement has been made in disease diagnostic techniques, with the most common methods used for identification or detection of infectious agents being microbial culture and colony counting, immunological assays (immunoassays) and molecular methods (Yang et al., 2008; Bissonnette and Bergeron, 2010; Kaittanis et al., 2010; Tallury et al. 2010).

*Corresponding Author Email: Pradipk14@yahoo.co.in.

CONVENTIONAL METHODS USED FOR PATHOGEN DETECTION

Microbial Culture and Colony Counting Method

Microbial culture and colony counting method is probably one of the oldest standard microbiological techniques of bacteria detection (Lazcka et al., 2007). It involves isolation of microorganism on a specific growth medium and identification of particular species, on the basis of color, shape and size of the colony formed (Lazcka et al., 2007; Tallury et al. 2010). However, the method has several limitations:

1) Selective medium is required for growing a particular species of the pathogen.
2) It is time-consuming, as an incubation period of 24 hours to 5 days is required for most of the pathogens (Lazcka et al., 2007; Tallury et al. 2010).
3) There is a requirement of extensive protocols to culture and grow pathogenic viruses in the laboratory.
4) It requires a typical laboratory setting, intensive sample handling, and preparation, and it cannot be utilized for onsite monitoring (Lazcka et al., 2007; Yang et al., 2008; Bissonnette and Bergeron, 2010; Hauck et al., 2010; Tallury et al. 2010).

Biochemical Tests

Biochemical tests involve the identification of pathogens by detecting a biochemical, enzymatic or metabolic product like alcohol, acids etc. The major limitation of the method is that it is labor-intensive and time-consuming (Bissonnette and Bergeron, 2010; Tallury et al. 2010).

Immunological Assays or Immunoassays

Immunological assays or immunoassays involve specific binding of an antibody to its antigen. In recent years, several immunological assays have been developed to detect bacteria, viruses and biotoxins. Some of the most widely-used immunoassays are agglutinations, radioimmunoassays, direct, competitive and sandwich ELISA (Enzyme Linked Immuno Sorbent Assay), immune-precipitation assay, enzyme-linked fluorescent assay (ELFA), and western blot test etc.(Lazcka et al., 2007; Yang et al., 2008; Tallury et al. 2010). Immunological assays have been widely employed for detecting food and waterborne pathogens. These methods are robust and sensitive, but the limitations include high cost of biological reagents, requirement of sophisticated and expensive instruments and lack of ability to detect pathogens in "real time" (Tallury et al. 2010; Velusamya et al., 2010).

Molecular Diagnostic Methods

Molecular diagnostic methods involve amplification of target DNA sequence of a pathogen (it can detect even a single copy of a DNA sequence) by PCR (Polymerase Chain Reaction). Different types of PCR, such as Reverse Transcriptase PCR (RT-PCR), Multiplex PCR, and Real time PCR have been used for the detection of pathogens (Deisingh and Thompson, 2004; Lazcka et al., 2007; Yang et al., 2008; Tallury et al. 2010). Using multiplex PCR, one can simultaneously detect several types of bacteria. However, the limitations of the method are that it cannot distinguish between viable and non-viable cells and false positive results may also be generated. In addition to the requirement of expensive reagents and sophisticated instruments, technically well-trained personnel are required to carry out the experiment. One of the major limitations of the method is that it scarcely lends itself to field experiments (Hauck et al., 2010).

There is an ever-increasing demand for advancement in the technology employed for pathogen detection to prevent morbidity and mortality caused by infectious diseases. Conventional methods used for pathogen detection have several limitations. The new technology or method for pathogen detection has to be accurate, specific, rapid, easy-to-use, economical and highly sensitive. Plenty of work has been done in the past two decades to develop robust, low-cost, easy-to-use, ultrasensitive and point-of-care diagnostic devices (Liu Y-T, 2008). In the last decade or so, several laboratories have developed novel methods for detection of microorganisms in food, water, complex mixtures (blood) and environment using nanotechnology (Tang, 2009).

Nanotechnology-Based Detection

Nanotechnology offers an enormous opportunity and unprecedented advantage in the development of rapid, easy-to-use, cost-effective and accurate diagnostics for pathogen detection (Jain, 2005; Rosi and Mirkin, 2005; Kaittanis et al., 2010). The field of nanotechnology has attracted such attention in the present time because the properties of nanomaterials (between 1.0 and 100 nm) are different from and more useful than those with higher particle size (Allaker and Ren, 2008). Nanotechnology involves the designing, creation or manipulation of structures at the nanometer level. Nanomaterials have large surface area, resulting in enhanced electrical and magnetic properties, and more intense quantum confinement effects and surface reactivity, (Gordon et al., 2007; Kaittanis et al., 2010; Blecher et al., 2011).

Nanotechnology has potential application in various fields such as physics, chemistry, biotechnology, energy production, microbiology, computer, electronics, medicine and molecular biology. In the recent times, world has witnessed many changes in diagnostic techniques with the introduction of nanotechnology and development of methods to engineer biofunctionalized nano-materials according to individual need. Nanomaterials offer many technological advantages over conventional methods used for the detection of pathogenic microorganisms (Jain, 2005; Rosi and Mirkin, 2005; Kaittanis et al., 2010).

Nano-biosensors are being developed using nanoparticles for rapid and accurate pathogen detection, with most of these methods being in the process of commercialization (Sanvicens et al., 2009). With the latest technologies available, properties of nanoparticles can be tailor-made according to specific need, just by changing the composition size and shape of the nanoparticle.

Nano-biosensor is an analytical device that has nanoscale detector components (Choi, 2012). It has three main elements: a bio-receptor or the biological recognition (sensitive) element (probe), the transducer, and the detector (Rai et al., 2012). The biological recognition element recognizes the analyte. Most widely used bio-receptors are antibodies, enzymes, nucleic acids, non-enzymatic protein, cell, and even the entire microorganism. The transducer plays an important role in signal detection and converts the biochemical signal into an electrical one (Sagadevan and Periasamy, 2014). Biological recognition element receives signals from the specific binding of bio-receptor to the analyte, generating a signal; the transducer then converts this (bio) chemical reaction or biological response into a measurable and quantifiable electrical output. Finally, the signal from the transducer is amplified, analyzed and the result (data) is displayed or stored (Rai et al., 2012; Sagadevan and Periasamy, 2014).Nanoparticles can be used in the receptor or transducer to enhance the sensitivity and performance of the nano-biosensor. In recent years, functionalized nanoparticles that can specifically bind to biomolecues (e.g., receptors, antibodies, enzymes and nucleic acids) have been developed to be used in nano-biosensors for microbial detection (Kaittanis et al., 2010).

Nanoparticles/Nanomaterials Used in Sensors for Detection

Nanoparticle-based biosensors are gaining enormous research attention. The use of nanomaterials in biosensors can improve their sensitivity and performance (Suresh and Mathan, 2014). Nanoparticles are particles of diameter less than 100 nm. Mostly, nanoparticles are hydrophobic. In order to use them for biological systems, solubilization and functionalization of nanoparticles is required. This renders them compatible with biological systems. They can be conjugated with different types of biomolecules (e.g., peptides, proteins, enzymes antibodies, nucleic acids and carbohydrates) (Luo and Stutzenberger, 2008). Nanoparticles can be organic (e.g., carbon nanotubes, liposomes, dendrimers etc.) or inorganic (quantum dots, gold nanoparticles, silver nanoparticles etc.). Nanoparticles like quantum dots, magnetic nanoparticles, gold nanoparticles, Eu-doped silica nanoparticles, liposomes, carbon nanotubes, and silica nanoparticles, etc. have been successfully used in the detection of pathogenic bacteria by several researchers Table 1.

There are several advantages of using nanoparticles in nano-biosensors:

1) Nanoparticles have extremely high surface area (i.e., large surface to volume ratio) due to their small size (1-100 nm). This property helps them interact better with the analyte (nanoparticles can bind to more number of biological molecules) (Kaittanis et al., 2010; Drbohlavova et al., 2009).
2) They have different or enhanced properties and functions, which significantly differ from those at the macro scale (nanoparticles have unique optical, electrical and magnetic properties) (Ghasemi et al., 2009; Kaittanis et al., 2010).
3) They can be easily synthesized or engineered according to the need by changing their composition, shape and size (Gooding, 2006; Kaittanis et al., 2010).

Nanoparticles have several advantages over conventional dyes; they are resistant to photo-bleaching, have narrow emission spectrum and high signal-to-noise ratio. The use of nanoparticles in biosensors will reduce the limit of detection (e.g., single cell can also be detected) and this, in turn, will reduce the sample size. In addition, different number and types of analyte/ pathogen can be detected in the same sample (Gooding, 2006; Lin et al., 2007).

Quantum Dots (QDs)

Quantum dots are fluorescent, spherical nanocrystals composed of inorganic semiconductors. The diameter of quantum dots ranges between 2-100 nm (Sweeney et al., 2008; Ghasemi et al., 2009). They have unique optical and electronic properties due to the quantum confinement effect. Quantum dot has an inorganic core of a semiconductor material. The core of QD is composed of group II and VI elements (e.g., CdSe and CdTe) or group III and V elements (Jamieson et al., 2007; Smith et al. 2008; Zrazhevskiy et al., 2010). The core is covered with an inorganic shell (ZnS) and a coating of an aqueous organic material for conjugating biomaterials such as antibodies, peptides or nucleotides (Alivisatos et al., 2005; Jain, 2005; Azzazy et al., 2007; Ghasemi et al., 2009). Quantum dots have many advantages over conventional fluorophores. They are resistant to photo-bleaching (i.e., they are stable) and exhibit unique luminescence property that allows them to be used as tags in imaging (Iga et al., 2007; Jamieson et al., 2007; Smith et al., 2008; Drbohlavova et al., 2009; Ghasemi et al., 2009) Multicolor quantum dots can be used to simultaneously detect multiple targets at the molecular level (True and Gao, 2007). They have exceptionally broad absorption spectra and narrow emission spectra.

Table 1. Nanoparticles (NPs) Used in pathogen detection

S. No.	Nanoparticle/ nanomaterial	Pathogen(s) detected	Limit of Detection/ Efficiency	Detection method	Reference
1	Quantum Dots	*E. coli* O157:H7	10^4 bacteria/ml	Fluorometry	Mukhopadhyay et al., 2009
2	Quantum Dots	Multiplexed E. coli, Salmonella	10^4CFU/ml	Fluorescence spectroscopy	Yang et al., 2006.
3	Quantum Dots	Cholera toxin, ricin, Shinga-like toxin1, *Staphylococcal* enterotoxin B	In ng/ml quantity	Fluoroimmunoassay	Goldman et al., 2004.
4	Magnetic NPs	*E. coli* O157:H7	10^4–10^5CFU/ml	IR Spectroscopy	Ravindranath et al., 2009.
5	Magnetic NPs	*E. coli* O157:H7	20 CFU/ml	ATP bioluminescence	Cheng et al., 2009.
6	Magnetic NPs	*Staphylococcus*	10 CFU/ml in human blood	TEM	Gao et al., 2006.
7	Magnetic NPs	*P fimbriated E. coli*	~9.6×10^4CFU/ 0.5 ml	MALDI-MS	Liu et al., 2008.
8	Magnetic NPs	*Listeria monocytogenes*	226 CFU/.5 ml	PCR	Yang et al., 2007.
9	Magnetic NPs	H5N1 virus	5 pg/ml	SQUID	Yang et al., 2008.
10	Magnetic NPs	*S. saprophyticus*	7×10^4CFU/ ml in urine	MALDI-MS	Liu et al., 2008.
11	Magnetic NPs	*S. aureus*	7×10^4CFU/ ml in urine	MALDI-MS	Liu et al., 2008.
12	Magnetic NPs	Gram-negative bacteria	4 CFU/ml	Fluorescence microscopy	Gu et al., 2003.
13	Magnetic NPs	*Mycobacterium tuberculosis*	20 colony-forming unit (CFU) in sputum (1 mL)	Nuclear Magnetic Resonance (NMR)	Lee et al., 2009.
14	Magnetic NPs	*S. aureus*	1 x 10^3CFU/mL	Magnetic Biosensor	Lee et al., 2008.
15	Superparamagnetic NPs	*E. coli* O157:H7	1x10^5CFU/mL	Magnetic Biosensor	Mujika et al., 2009.
16	Superparamagnetic ironoxide NPs	*Mycobacterium avium* spp. paratuberculosis (MAP),	38.75 CFUs	Nanosenor(Magnetic relaxation)	Kaittanis et al., 2007

S. No.	Nanoparticle/ nanomaterial	Pathogen(s) detected	Limit of Detection/ Efficiency	Detection method	Reference
17	Gold NPs	*S. Enteritidis* insertion-element gene	1 ng/mL	Optical biosensor	Zhang et al., 2009.
18	Gold NPs	*B. subtilis* genomic DNA	2.5 fM	Optical biosensor	Hill et al., 2007.
19	Gold NPs	*S. aureus*	7 · 105CFU/mL	Surface plasmon resonance biosensor	Joung et al., 2008.
20	Gold NPs	*E. coli*	58.2 ± 1.37 pg/m	Surface plasmon resonance biosensor	Joung et al., 2008.
21	Gold NPs	*E. coli* O157:H7	10 ng	Microscope and visual	Lin et al., 2005.
22	Gold NPs	Cholera toxin	10 fg/ml in 20 min	Test strip	Ahn-Yoon et al., 2003.
23	Gold NPs	*H. pylori*	10 ng	SEM	Lin et al., 2005.
24	Gold nanowire arrays	*E. coli*	10 cells/0.173 cm2	Electrochemical Impedance spectroscopy	Basu, 2004.
25	Gold and Magnetic NPs	*Salmonella enteritidis*	1 ng/ml	Fluorescence	Zhang et al., 2009.
26	Gold NPs/Ag staining	*Giardia lamblia*	1.088×10^{3}cells/ml	UV–Vis spectroscopy	Li et al., 2009.
27	Gold NPs	*A. azurea, B. Subtilis, E. coli, P. putida*	2×105 cells/mL	array-based sensing	Phillips et al., 2008
28	Gold nanorods	*Escherichia coli*	50 colony forming unit (CFU)/mL	optical sensors (two-photonscattering technique)	Singh et al., 2009
29	Eu doped polystyrene NPs	Adenovirus	800 folds more sensitive thanconventional immunoassays	Fluoroimmunoassay	Valanne et al., 2005.
30	Eu doped silica NPs	*B. anthracis*	0.2 nM in 2 min	Fluorescence spectra	Kelong et al., 2009.
31	Eu doped polystyrene NPs	*B. anthracis*	0.01–100 ng/ml	Fluoroimmunoassay	Tang et al., 2009.
32	Liposomes	Cholera toxin	1 nM	Fluoroimmunoassay	Singh et al., 2000.
33	Liposomes	Influenza virus	1 HAUs	Colorimetry	Reichert et al., 1995.
34	Liposome	*E. coli* O157:H7	1.0×10^{2} CFU/ml	immunoassay	Ho et al., 2003.
35	Carbon nanotubes	*E. coli*	10^{4} CFU/ml	Dielectrophoresis (DEP)	Zhou et al., 2006.
36	Carbon magnetic nanotubes	*E. coli* O157:H7	40 CFU/0.1 ml	IMS	Lin et al., 2006.
37	Single-walled carbon nantotubes	*S. infantis*	100 CFU/mL	Electrochemical biosensor	Villamizaret al., 2008.
38	Silica nanoparticles	*E. coli* O157:H7	Single Bacterium	Spectrofluorometric analysis	Zhao et al., 2004.
39	Silica coated silver (Ag@SiO2)	*E. coli* O157:H7	13 cfu/mL	Transmission Electron Microscope (TEM) & UV-vis absorption	Chen et al., 2014
40	Graphene oxide	*L. acidophilus*	~11.0 cfu mL−1	nano-biosensors	Zuo et al., 2013

Gold Nanoparticles

Gold nanoparticles with a diameter of 1-100 nm are available in several forms (Gold nanorods, nanoshells, nanofibers and nanocages). Gold nanoparticles (AuNP) have high surface-to-volume ratio. They have unique size and shape-dependent, tunable, optical, electrochemical and magnetic properties. They also possess unique surface plasmon property (Yigit and Medarova, 2012). Gold nanoparticles have been extensively used in biosensors for sensing and detecting pathogens (Allaker and Ren, 2008). Their surface can be easily modified and, therefore, they are widely used in different chemical and biological sensing systems for molecular recognition. Gold nanowires are one-dimensional nanostructures widely used in nanosensing devices for biochemical applications (Fendler, 1996; Lal et al., 2006; Zhou et al., 2009; Cobley et al., 2011; Dawson and O'Riordan, 2011; Yang et al., 2011, Zhang et al., 2014).

Silver Nanoparticles

Silver nanoparticles of 1-100nm size have unique physicochemical (optical and electrical conductivity, and thermal properties) and biological properties, such as high catalytic activity and chemical stability. Silver nanoparticles are used as biological tags in molecular detection. Their unique properties are shape and size-dependent (Krutyakov et al., 2008; Habouti et al., 2010).

Nanotechnology-Based Detection of Microorganisms

According to CDC (Centers for Disease Control), *Escherichia coli* O157:H7, *Salmonella* spp. and *L. monocytogenes* are amongst the seven pathogens that are leading causes (90%) of foodborne illnesses, hospitalizations, and deaths due to known pathogens (http://www.cdc.gov/foodborneburden/questions-and-answers.html). Chen and Durst (2006) reported that protein G (obtained from group G streptococci)-liposomal nanovesicles can be used as effective universal reagents in both direct and competitive immunoassays. They developed an array-based immunosorbent assay using these nanovesicles for simultaneously detecting *E. coli* O157:H7, *Salmonella enteric* serovar Typhimurium, and *L. monocytogenes* in pure and mixed cultures. The protein G-liposomal nanovesicles contained millions of fluorescent dye molecules that provided improved and strong signals in the assay. The detection limit of the direct competitive immunoassay for pure culture was approximately 100 cfu/ml for *E. coli* O157:H7, 500 cfu/ml for *Salmonella enteric* serovar Typhimurium and $1.5x10^4$ cfu/ml for *Listeria monocytogenes.* The detection limit was approximately 3.1×10^3 for *E. coli* O157:H7, 7.8×10^4 *for Salmonella enteric* serovar Typhimurium, and 7.9×10^5 CFU/ml *Listeria monocytogenes.* The sensitivity of this assay was comparable to Standard ELISA methods.

Shigella flexneri is the causative organism of acute bloody diarrhea termed shigellosis. Li and Xu (2009) reported an ultrasensitive method for the detection of *Shigella flexineri* using antibody-coated fluorescent silica nanoparticles (~50 nm). Nanoparticles were resistant to photo-bleaching. They provided a strong and reproducible signal and long microscopic observation of the target bacteria.

Kim et al. (2007) reported that the use of antibody-conjugated nanoparticles (streptavidin-coated quantum dots) on the impedimetric biosensor surface significantly enhanced the detection limit of *Salmonella entriditis* in PBS (LOD 10^4cfu/ml) and milk (LOD 10^5cfu/ml).

Gu et al. (2003) employed bifunctional magnetic nanoparticles (diameter<10 nm) for rapid and sensitive detection of vancomycin-resistant enterococci (VRE). The vancomycin-FePt conjugate used polyvalent interactions to capture VRE at a concentration of 10^1-10^2cfu/ml in an hour. These magnetic nanoparticles also captured and detected other Gram positive bacteria like *Staphylococcus aureus*, *S. epidermidis*, and a coagulase-negative staphylococci (CNS) at a concentration of 8 cfu/mL, 10 cfu/mL, and 4 cfu/mL, respectively. To capture and detect these bacteria at such ultralow concentration, use of polyvalent magnetic nanoparticles (3-4 nm) in comparison to magnetic beads (1-5 μm) showed high performance due to their large surface-to-volume ratio.

Fourier Transform Infrared (FTIR) spectroscopy has been used extensively in pathogen detection (bacteria). With the introduction of nanoparticles in biosensors, this technique is finding more applications, especially in nanobiosensors.

Mura et al. (2012) developed an optical biosensor for the selective and specific detection of *E. Coli* O157:H7 by using FTIR spectroscopy. They used mesoporoustitania thin films, functionalized with APTES ((3-amino-propyl) triethoxysilane) and GA (glutaraldehyde), to immobilize anti *E.coli* O157: H7 monoclonal antibody. This mesoporoustitania thin film-based FTIR nanobiosensor showed a detection limit of 1×10^2 CFU/mL and the bacteria was captured in less than 30 minutes. This simple new biosensor (nanodevice) has the potential to be used for screening water sample contaminated with *E. coli* O157:H7.

Legionella pneumophila is a pathogenic bacterium, which causes Legionnaires' disease (Reingold et al., 1984; Yu et al., 2002). Rapid diagnosis of this disease is necessary for proper treatment and survival of the patient. Rai et al. (2012) developed a "nanoporous alumina membrane-based label-free electrochemical DNA biosensor" for the detection of *L. pneumophila.* They detected a 21-mer complementary analyte DNA sequence of *L. pneumophila*. The major highlights of their biosensor was that it showed ultrasensitive detection of *L. pneumophilla* DNA with a Limit of Detection (LOD) of 3.1×10^{-13} M. Also, this biosensor could differentiate one and three base-pair mismatches of the complementary sequence from the target sequences of different strains of *Legionella* sp.

Electrochemical nanoporous alumina membrane-based DNA biosensors can have multiple applications in various fields, as in addition to being portable and cost-effective, they can detect the genomic sequence of the pathogenic microorganism (Rai et al., 2012).

Dengue is a viral disease, transmitted by mosquitoes (genus *Aedes*) infected with any one of the four distinct serotypes of the dengue virus (DEN-1, DEN-2, DEN-3, and DEN-4). The dengue virus causes illness, sudden fever, acute joint pains and death in many parts of the world (http://www.who.int/denguecontrol/en/) (http://www.cdc.gov/dengue/). Cheng et al. (2012) employed a membrane-based electrochemical nanobiosensor that recognized dengue type -2 viruses (DENV-2) against two other non-specific arthropod-borne viruses, namely the Chickengunya virus (CHIKV), and the West Nile Virus (WNV) and another serotype of dengue virus (DENV-3). IgG (monoclonal antibody) raised against DENV-2 was used as the specific bio-recognition element for detection of DENV-2. Major highlight of this nanobiosensor was that it was ultrasensitive and specific in the detection of dengue type -2 viruses (DENV-2), having a remarkably low detection limit (1 pfu/ml) with linear detection range from $1\text{-}10^3$pfu/ml; the analysis time was also short (ca. 50 minutes) as compared to other methods of identification. This nanobiosensor might be used for diagnosis of dengue at point-of-care and for selective identification and quantification of dengue type -2 viruses (DENV-2) in *Aedesaegypti* mosquito in the field.

West Nile Virus (WNV) can cause serious, sometimes fatal neurological disease and death in humans. It is usually transmitted to humans by infected mosquitoes. (http://www.who.int/mediacentre/factsheets/fs354/en/)(http://www.cdc.gov/westnile/index.html). Nguyen et al. (2009) fabricated a 'membrane-based electrochemical nanobiosensor' to detect whole virus particle or virus E protein of the West Nile Virus (WNV) by forming a "nanoporous alumina membrane over a sensing electrode (platinum disk sensing electrode)."Anti-WNV monoclonal antibody (IgM) was used as the bio-recognition element for the detection of WNV particle. The limit of detection of this assay was extremely low (4pg/ml) and ca. 2 viral particles/100 ml, which was comparable to the sensitivities of PCR techniques and the detection time was very short, i.e., 30 minutes. In future, this nanobiosensor may be suitable for detecting WNV virus in an early stage of infection.

Zhao et al. (2004) reported antibody conjugated RuBpy doped silica nanoparticles based immunoassay for in situ identification and quantification of pathogens. The silica nanoparticle-based bioassay could detect a single bacterial cell (*E. coli* O157:H7) per given sample in less than 20 minutes. They bioconjugated monoclonal antibodies against *E. coli* O157:H7. Dye-doped silica nanoparticles with anti *E. coli* O 157:H7 were added to *E. coli* O157-inoculated ground beef. The method was quick, ultrasensitive and accurate as compared to conventional methods of bacterial detection. There was no requirement of any amplification or enrichment step for bacterial detection. Antibody-conjugated silica nanoparticles (diameter of 60+4nm) bound to *E. coli* O157:H7 cells were detected by flow cytometer. Thousands of bioconjugated nanoparticles can bind to many surface antigens available for specific recognition on a bacterium. Each silica nanoparticle encapsulated thousands of dye molecules and this enables high signal amplification of the antibody-conjugated RuBpy-doped silica nanoparticles.

CONCLUSION

Nanotechnology has shown significant potential in developing devices and methods for disease diagnosis. The biggest advantage of nanotechnology in the field of diagnostics is that there is no need to prepare the sample for analysis, and automation is easily possible. The use of nanoparticles in detection and diagnosis of pathogens in food, clinical samples and complex media such as milk has shown promising results towards the development of nanotechnology-based diagnostics with improved features. However, significant development is required to improve the current technology. Advancements in the field of nanotechnology have contributed to the miniaturization of the devices. In future, an ideal nano-biosensor will have the desirable attributes of high sensitivity, high selectivity, multiplexing ability, portability, ease of handling and economical feasibility. Laboratories across the globe are working on the development of portable, easy-to-handle devices with a high degree of sensitivity [Sanvicens et al., 2009]. Despite rapid advancement in the field of nanotechnology, there are many future challenges that need to be addressed.

REFERENCES

Ahn-Yoon S, DeCory TR, Baeumner AJ, Durst RA. Ganglioside–liposome immunoassay for the ultrasensitive detection of cholera toxin. *Anal. Chem.* 2003; 75:2256–2261.

Alivisatos AP, Gu W, Larabell C. Quantum dots as cellular probes. *Annu Rev Biomed Eng* 2005;7:55–76.

Allaker RP, Ren G. Potential impact of nanotechnology on the control of infectious diseases. *Transact Roy Soc Trop Med Hyg*. 2008; 102: 1-2.

Azzazy HME, Mansour MMH, Kazmierczak SC. From diagnostics to therapy: Prospects of quantum dots. *Clin Biochem.* 2007;40:917–927.

Basu SM. Nano-biosensor development for bacterial detection during human kidney infection: use of glycoconjugate-specific antibody-bound gold NanoWire arrays (GNWA). *Glycoconj J*. 2004; 21:487–496.

Bissonnette L, Bergeron MG. Diagnosing infections—current and anticipated technologies for point-of-care diagnostics and home-based testing. *ClinMicrobiol Infect*. 2010;16:1044–1053.

Blecher K, Nasir A, Friedman A. The growing role of nanotechnology in combating infectious disease. *Virulence*. 2011; 2(5): 395-401.

Chen CS, Durst RA. Simultaneous detection of *Escherichia coli* O157:H7, *Salmonella* spp. And *Listeria monocytogenes* with an array-based immunosorbent assay using universal protein G-liposomal nanovesicles.*Talanta*. 2006; 69:232–238.

Chen GZ, Yin ZZ, Lou JF. Electrochemical Immunoassay of Escherichia coli O157:H7 Using Ag@SiO2 Nanoparticles as Labels. *J Analyt Meth Chem*. 2014: 7

Cheng MS, Ho JS, Tan CH, Wong JPS, Ng LC, Toh CS. Development of an electrochemical membrane-based nano biosensor for ultra sensitive detection of dengue virus. *Anal ChimActa*. 2012;725:74– 80.

Cheng Y, Liu Y, Huang J, Li K, Zhang W, Xian Y, Jin L. Combining biofunctional magnetic nanoparticles and ATP bioluminescence for rapid detection of Escherichia coli. *Talanta*. 2009:77: 1332–1336.

Choi C. Integrated nanobiosensor technology for biomedical application. *Nanobiosens. Dis. Diag*. 2012;1:1–4.

Cobley CM, Chen J, Cho EC, Wang LV, Xia Y. Gold nanostructures: A class of multifunctional materials for biomedical applications. *Chem. Soc. Rev*. 2011; 40: 44–56.

Dawson K, O'Riordan A. Towards nanowire (bio) sensors. *J. Phys. Conf. Ser*. 2011; 307.

Deisingh AK, Thompson M. Biosensors for the detection of bacteria. *Can J Microbiol*. 2004;50(2):69–77.

Drbohlavova J, Adam V, Kizek R, Hubalek J. Quantum Dots — Characterization, Preparation and Usage in Biological Systems. *Int. J. Mol. Sci*. 2009;10:656-673.

Fendler JH. Self-assembled nanostructured materials. *Chem. Mater*. 1996; 8: 1616–1624.

Gao J, Li LH, Ho PL, Mak GC, Gu HW, Xu B. Combining fluorescent probes and biofunctional magnetic nanoparticles for rapid detection of bacteria in human blood. *Adv. Mater*. 2006; 18: 3145–3148.

GhasemiY, Peymani P, Afifi S. Quantum dot: magic nanoparticle for imaging, detection and targeting. *Acta Biomed*. 2009; 80: 156-165

Goldman ER, Clapp AR, Anderson GP, Uyeda HT, Mauro JM, Medintz IL, Mattoussi H. Multiplexed toxin analysis using four colors of quantum dot fluororeagents. *Anal. Chem*. 2004;76: 684–688.

Gooding JJ. Nanoscale Biosensors: Significant Advantages over Larger Devices? *Small*. 2006; 2(3):313 – 315.

Gordon AT, Lutz GE, Cooper RA, Boninger ML: Introduction to nanotechnology: potential applications in physical medicine and rehabilitation. *Am. J. Phys. Med. Rehabil.*2007;86:225–241.

Gu H, Ho PL, Tsang KWT, Wang L, Xu B. Using Biofunctional Magnetic Nanoparticles to Capture Vancomycin-Resistant Enterococci and Other Gram-Positive Bacteria at Ultralow Concentration. *J. Am. Chem. Soc.* 2003;125:15702-15703.

Gu HW, Ho PL, Tsang KWT, Yu CW, Xu B. Using biofunctional magnetic nanoparticles to capture Gram-negative bacteria at an ultra-low concentration. *Chem. Commun.* 2003;1966–1967.

Habouti S. Solterbeck CH, Es-Souni M. Synthesis of silver nano-fir-twigs and application to singlemolecules detection. *J. Mater. Chem.* 2010; 20: 5215–5219.

Hauck TS, Giri S,Gao Y, Chan WCW. Nanotechnology diagnostics for infectious diseases prevalent in developing countries. *Adv. Drug Deliv. Rev.* 2010;62:438–448.

Hill HD, Vega RA, Mirkin CA. Nonenzymatic detection of bacterial genomic DNA using the biobar code assay. *Anal. Chem.* 2007;79 (23): 9218-9223.

Ho JA, Hsu HW. Procedures for preparing Escherichia coli O157:H7 immunoliposome and its application in liposome immunoassay. *Anal Chem.* 2003; 75:4330–4334

http://www.cdc.gov/foodborneburden/questions-and-answers.html.

http://www.who.int/denguecontrol/en/) (http://www.cdc.gov/dengue/.

http://www.who.int/mediacentre/factsheets/fs354/en/)(http://www.cdc.gov/westnile/index .html.

Iga AM, Robertson JHP, Winslet MC, Seifalian AM. Clinical potential of quantum dots. *J. Biomed. Biotech.* 2007;1-10.

Jain KK. Nanotechnology in clinical laboratory diagnostics. *Clin. Chim. Acta.* 2005;358:37–54.

Jamieson T, Bakhshi R, Petrova D, Pocock R, Imani M, Seifalian AM. Biological applications of quantum dots. *Biomaterials.* 2007;28: 4717–4732.

Joung HA, Lee NR, Lee SK, Ahn J, Shin YB, Choi HS, Lee CS, Kim S, Kim MG. High sensitivity detection of 16s rRNA using peptide nucleic acid probes and a surface plasmon resonance biosensor. *Anal. Chim. Acta.* 2008; 630(2): 168-173.

Kaittanis C, Naser SA, Perez JM. One-step, nanoparticle-mediated bacterial detection with magnetic relaxation. *Nano. Lett.* 2007;7:380–383.

Kaittanis C, Santra S, Perez JM. Emerging nanotechnology based stratergies for the identification of microbial pathogenesis. *Adv Drug Deliv Rev.* 2010;62(4-5):408–423.

Kelong Ai BZLL, Europium-based fluorescence nanoparticle sensor for rapid and ultrasensitive detection of an anthrax biomarker13. *Angew. Chem., Int. Ed.* 2009;48: 304–308.

Kim G, Mun JH, Om AS. Nano-particle enhanced impedimetric biosensor for detection of food borne pathogens. *Journal of Physics: Conference Series.*2007;61: 555–559.

Krutyakov YA, Kudrynskiy AA, Olenin AY, Lisichkin GV.Synthesis and properties of silver nanotpaticles: Advances and prospects. *Russ. Chem. Rev.* 2008;77:233-257.

Lal M, Plummer M, Smith W. Solvent density effects on the solvation behavior and configurational structure of bare and passivated 38-atom gold nanoparticle in supercritical ethane. *J. Phys. Chem. B* 2006; 110: 20879–20888.

Lazcka O, Campo FJD, Munoz FX. Pathogen detection: a perspective of traditional methods and biosensors. *Biosen. Bioelec.* 2007;22:1205-1217.

Lee H, Sun E, Ham D, Weissleder R. Chop-NMR biosensor for detection and molecular analysis of cells. *Nat. Med.* 2008; 14: 869.

Lee H, Yoon TJ, Ralph Weissleder R. Ultrasensitive detection of bacteria using core-shell nanoparticles and a NMR-filter system. *Angew Chem Int Ed Engl.* 2009; 48(31): 5657–5660.

Li XX, Cao C, Han SJ, Sim SJ. Detection of pathogen based on the catalytic growth of gold nanocrystals. *Water Res.* 2009; 43: 1425–1431.

Li Y, Xu W. Highly sensitive detection of *Shigellaflexneri* using fluorescent silica nanoparticles. *New Microbiologica.* 2009; 32:377-383.

Lin FYH, Sabri M, Alirezaie J, Li D, Sherman PM. Development of a nanoparticle-labeled microfluidic immunoassay for detection of pathogenic microorganisms. *Clin. Diagn. Lab. Immunol.* 2005; 12: 418–425.

Lin S, Xie X, Patel MK, Yang YH, Li Z, Cao F, Gheysens O, Zhang Y, Gambhir SS, Rao JH, Wu JC. Quantum dot imaging for embryonic stem cells. *BMC Biotechnology.* 2007;7:67.

Lin Y, Jiang XP, Elkin T, Fernando KAS, Gu LR, Taylor S, Yang H, Jones E, Wang W, Sun YP.Carbon nanotubes for immunomagnetic separation of Escherichia coli O157:H7. *J. Nanosci Nanotechnol.* 2006;6(3):868–887.

Liu JC, Tsai PJ, Lee YC, Chen YC. Affinity capture of uropathogenic Escherichia coli using pigeon ovalbumin-bound Fe_3O_4@Al_2O_3 magnetic nanoparticles, *Anal. Chem.* 2008;80: 5425–5432.

Liu YT. A technological update of molecular diagnostics for infectious diseases. *Infect Disord Drug Targets.* 2008; 8(3): 183–188.

Luo P, Stutzenberger F. Nanotechnology in the detection and control of microorganisms. *Adv. Appl. Microbiol.* 2008;63:145-181.

Mujika M, Arana S, Castao E,Tijero M, Vilares R, Ruano-Lo pez JM, Cruz A, Sainz L, Berganza J. Magnetoresistiveimmunosensor for the detection of Escherichia coli O157:H7 including a microfluidic network. *Biosens. Bioelectron.* 2009; 24(5): 1253-1258.

Mukhopadhyay B, Martins MB, Karamanska R, Russell DA, Field RA. Bacterial detection using carbohydrate-functionalisedCdS quantum dots: a model study exploiting E. coli recognition of mannosides. *Tetrahedron Lett.* 2009; 50: 886–889.

Mura S, Greppi G, Marongiu ML, Roggero PP, Ravindranath SP, Mauer LJ, Schibeci N, Perria F, Piccinini M, Innocenzi P, Irudayaraj J. FTIR nanobiosensors for Escherichia coli detection. *Beilstein J. Nanotechnol.* 2012;3:485–492.

Nguyen BT, Koh G, Lim HS, Chua AJ, Ng MM, Toh CS. Membrane-based electrochemical nanobiosensor for the detection of virus. *Anal Chem.* 2009; 81:7226-7234.

Phillips RL, Miranda OR, You CC, Rotello VM, Bunz UHF. Rapid and efficient identification of bacteria using gold-nanoparticle - Poly(para-phenyleneethynylene) constructs. *Angew Chem Int Ed.* 2008;47:2590–2594.

Rai V, Acharya S, Dey N. Implications of Nanobiosensors in Agriculture. *J. Biomat. Nanobiotech.* 2012;3:315-324.

Rai V, Deng J, Toh CS. Electrochemical nanoporous alumina membrane-based label-free DNA biosensor for the detection of Legionella sp. *Talanta.* 2012; 98:112–117.

Ravindranath SP, Mauer LJ, Deb-Roy C, Irudayaraj J. Biofunctionalized magnetic nanoparticle integrated mid-infrared pathogen sensor for food matrixes. *Anal. Chem.* 2009;81: 2840–2846.

Reichert A, Nagy JO, Spevak W, Charych D. Polydiacetylene liposomes functionalized with sialic acid bind and colorimetrically detect influenza virus. *J. Am. Chem. Soc.* 1995;117: 829–830.

Reingold AL, Thomason BM, Brake BJ. *J. Infect. Dis.* 1984;149:819.

Rosi NL, Mirkin CA. Nanostructures in biodiagnostics. *Chem. Rev.* 2005;105:1547–1562.

Sagadevan S, Periasamy M. Recent trends in nanobiosensors and their applications - A review. *Rev. Adv. Mater. Sci.* 2014;36:62-69.

Sanvicens N, Pastells C, Pascual N, Marco MP. Nanoparticle-based biosensors for detection of pathogenic bacteria, *Trends Anal Chem.* 2009; 28(11):1243-1252.

Singh AK, Harrison SH, Schoeniger JS. Gangliosides as receptors for biological toxins: development of sensitive fluoroimmunoassays using ganglioside-bearing liposomes, *Anal. Chem.* 2000; 72: 6019–6024.

Singh AK, Senapati D, Wang S. et al. Gold nanorod based selective identification of Escherichia coli bacteria using two photon rayleigh scattering spectroscopy. *ACS Nano.* 2009; 3(7):1906–1912.

Smith AM, Duan H, Mohs AM, Nie S. Bioconjugated Quantum Dots for *In Vivo* Molecular and Cellular Imaging. *Adv Drug Deliv Rev.* 2008; 60(11): 1226–1240.

Suresh S, Mathan P. Recent Trends In Nanobiosensors and Their Applications – A Review. *Rev. Adv. Mater. Sci.* 2014;36: 62-69.

Sweeney E, Ward TH, Gray N, Womackb C, Jayson G, Hughes A, Dive C, Byers R,Quantitative multiplexed quantum dot immunohistochemistry. *Biochem. Biophys Res Comm.* 2008; 374:181–186.

Tallury P, Malhotra A, Byrne LM, Santra S. Nanobioimaging and sensing of infectious disease. *Adv. Drug. Del. Rev.* 2010;52:424-437.

Tang S, Moayeri M, Chen Z, Harma H, Zhao J, Hu H, Purcell RH, Leppla SH, Hewlett IK. Detection of anthrax toxin by an ultrasensitive immunoassay using Europium nanoparticles. *Clin. Vaccine Immunol.* 2009;16:408–413.

Taubes G. The bacteria fight back. *Science.* 2008; 321:356–361.

True LD, Gao X. Quantum dots for molecular pathology their time has arrived. *J. Mol. Diag.* 2007;9:7-11.

Valanne A, Huopalahti S, Soukka T, Vainionpää R, Lövgren T, Härmä H. A sensitive adenovirus immunoassay as a model for using nanoparticle label technology in virus diagnostics, *J. Clin. Virol.* 2005;33:217–223.

Velusamya V, Arshaka K, Korostynskaa O, Oliwab K, Adleyb C. An overview of foodborne pathogen detection: In the perspective of biosensors. *Biotechnology Advances.* 2010; 28: 232–254.

Villamizar RA, Maroto A, Rius FX, Inza I, Figueras MJ,Fast detection of Salmonella infantis with carbon nanotubes field effect transistors. *Biosens. Bioelectron.* 2008;24: 279-283.

Wang C, Irudayaraj J.Gold Nanorod Probes for the Detection of Multiple Pathogens. *Small.* 2008;4(2):2204–2208.

Yang H, Qu LW, Lin Y, Jiang XP, Sun YP, Detection of Listeria monocytogenes in biofilms using immunonanoparticles, *J. Biomed. Nanotechnol.* 2007; 3: 131–138.

Yang L, Li Y, Simultaneous detection of Escherichia coli O157:H7 and Salmonella typhimurium using quantum dots as fluorescence labels. *Analyst.* 2006;131: 394–401.

Yang SY, Chieh JJ, Wang WC, Yu CY, Lan CB, Chen JH, Horng HE, Hong CY, Yang HC, Huang W. Ultra-highly sensitive and wash-free bio-detection of H5N1virusby immunomagnetic reduction assays, *J. Virol. Methods.* 2008; 153: 250–252.

Yang W, Qu L, Zheng R, Liu Z, Ratinac KR, Shen L, Yu D, Yang L, Barrow CJ, Ringer SP, et al. Self-assembly of gold nanowires along carbon nanotubes for ultrahigh-aspect-ratio hybrids. *Chem. Mater.* 2011; 23: 2760–2765.

Yigit MV, Medarova Z. *In vivo* and *ex vivo* applications of gold nanoparticles for biomedical SERS imaging. *Am. J. Nucl. Med. Mol. Imaging.* 2012;2(2):232-241.

Yu VL, Plouffe JF, Pastoris MC, Stout JE, Schousboe M, Widmer A, Summersgill J, File T, Heath CM, Paterson DL, Chereshsky A. *J. Infect. Dis.* 2002; 186: 127–128.

Zhang D, Carr DJ, Alocilja EC. Flourescentbiobarcode DNA assay for the detection of salmonella entericaserovarEnteritidis. *Biosens. Bioelectron.*2009; 24(5): 1377-1381.

Zhang Y, Chu W, Foroushani AD,Wang H, Li D, Liu J, Barrow CJ, Wang X, and Yang W. New Gold Nanostructures for Sensor Applications: A Review. *Materials.* 2014; 7: 5169-5201.

Zhao X, Hilliard LR, Mechery SJ, Wang Y, Bagwe RP, Jin S, Tan W. A rapid bioassay for singlebacterial cell quantitation using bioconjugated nanoparticles. *Proc. Natl. Acad. Sci. U S A.*2004;101:15027–15032.

Zhou R, Wang P, Chang HC. Bacteria capture, concentration and detection by alternating current dielectrophoresis and self-assembly of dispersed single-wall carbon nanotubes. *Electrophoresis.* 2006; 27:1376–1385.

Zhou J, Ralston J, Sedev R, Beattie DA. Functionalized gold nanoparticles: Synthesis, structure and colloid stability. *J. Colloid Interface Sci.* 2009, 331, 251–262.

Zrazhevskiy P, Sena M, Gao X. Designing multifunctional quantum dots for bioimaging, detection, and drug delivery. *Chem. Soc. Rev.* 2010 November; 39(11): 4326–4354.

Zuo P, Li XJ, Dominguez DC, Ye BC. A PDMS/paper/glass hybrid microfluidic biochip integrated with aptamer-functionalized graphene oxide nano-biosensors for one-step multiplexed pathogen detection. *Lab Chip.* 2013; 13(19): 3921–3928.

In: Recent Advances in Microbiology. Volume 3 ISBN: 978-1-53614-057-6
Editors: N.C. Gautam, S.P. Tiwari and R. Sharma

Chapter 11

AN OVERVIEW ON SOIL MICROORGANISMS PATHOGENIC TO VEGETABLE CROPS: DETECTION, DEVELOPMENT AND MANAGEMENT

Bineet Kumar Sharma, Rakesh Kumar Rai and Sujoy Saha*
ICAR-Indian Institute of Vegetable Research, Varanasi, India

INTRODUCTION

Soil contains a plethora of microorganisms, with both beneficial and deleterious attribute constituting an environmental surrounding termed as "soil micro-flora". Soil containing beneficial microbes is a boon for agricultural crops, but existence of harmful microorganisms can be detrimental to both plant and animal health. Diverse population of microorganisms like fungi, bacteria, protozoa, algae and actinomycetes constitute a specific soil micro-flora of a specific region.

Vegetables are the major component of nutritional and livelihood security. As the production of crops depends largely on soil micro-flora, soil rich with beneficial micro-flora/microorganism is known to enhance crop productivity. The physico-chemical nature of soil is also dependent upon the presence of microorganisms in the soil. The population of microorganism also reflects the fertility status of specific soil. Productivity of vegetables per unit area is diminishing and there is a necessity to focus on soil environment. Pathogenic soil microorganisms play a major role in decreasing vegetable production and effective management of them is a challenge. The present chapter deals

*Corresponding Author e-Mail:sujoyta@gmail.com.

with several soil microorganisms viz. *Fusarium* sp., *Verticillium* sp., *Pythium* sp., *Rhizoctonia* sp., *Sclerotium* sp., *Sclerotinia* sp. and *Ralstonia* sp. which are harmful to soil health causing several diseases like wilt, damping off, rot etc. on a wide variety of vegetable crops. An effort is also made to construct a roadmap for their effective control.

DETECTION

Microorganisms involved in causing soil borne diseases are not easy to detect because of their soil inhabiting nature; they mostly cause collar rot disease in vegetable crops which never shows any visible appearance like, blight, spot and mildew structure. The appearance of symptoms is usually localized post infection by the soil inhabiting microorganisms. Detection of any specific microorganism at the right time of the disease development is very crucial. Symptomatological, genetic, biochemical or molecular detection techniques for the detection of soil borne plant pathogenic microorganism have been reported.

Symptomatological Detection

It is the oldest but a reliable method of detection of plant pathogenic microorganism. Detection of different soil microorganism pathogenic to vegetable crops is discussed here:

Fusarium spp.- They are the common soil microorganisms found worldwide and notorious for causing wilt disease in a wide range of agricultural crops including vegetables. The initial symptom of *Fusarium* spp. is yellowing of the foliage, beginning with the lower leaves and progressing upwards. Sometimes yellowing can be seen in only one side of a leaf midrib, or branch or one side of the affected plant (Walker, 1957). At later stage vascular browning can be seen in infected stems. *Fusarium* is reported to produce three types of asexual spores *viz*- microconidia, macrocinidia and chlamydospores (Agrios, 2004).

Verticillium spp.- The symptom of *Verticillium* wilt may be confused with those of *Fusarium* wilt. As the symptoms on plants caused by these pathogens are similar, detection in the laboratory is important. In the later stages the entire plant Become wilted with green, yellow and brown discoloration (Douglas, 2008). Discolored vascular tissue may be observed when the stem is dissected.

Pythium spp.- Mostly seeds/seedlings are attacked by this pathogen and they fail to germinate. Emerged seedlings show water soaked area and discoloration at the collar region near the soil line. The lower part of seedling adjacent to the soil become thinner and result in drooping of the seedlings. *Pythium* produces white, rapidly growing mycelium, in the affected regions (Broders et al., 2009).

Rhizoctonia spp.- It causes wire stem disease in several vegetable crops. It also affects prior to seedling emergence. Young seedlings show soft, water soaked spot on the stem near collar region just like damping off.

Sclerotium spp.- It is also responsible for collar rot disease in several vegetable crops. The first symptom of the disease is observed as soft tissue necrosis on the bark of the infected stem near the collar region just beneath the soil. Infected portion is covered by white cottony and silver color mycelium. At the later stage mustard like "sclerotia" are seen on infected parts of the plant (Verma and Sharma, 1999).

Sclerotinia spp.- It causes "white rot" disease of different vegetable crops. The main symptomatological detection of this pathogen is presence of white/ fluffy cottony mycelium at the point of infection on plant. Later a compact resting body "sclerotia" can be produced.

Ralstonia spp.- Bacterial wilt disease in vegetable crops is caused by this bacteria. Pathogen is mostly confined to vascular region. In advanced stage, it may invade the cortex and pith and cause yellow brown discoloration in the infected tissues. An oozing at the cut end of the stem is a typical symptom of bacterial wilt disease (Verma and Sharma, 1999).

Genetic Detection

It includes sexual and vegetative compatibility methods for the detection of soil microorganism pathogenic to vegetable crops. These methods are dependent on inherent genetic characteristics of the related microorganism which are important tools for the purpose of identification.

Sexual compatibility denotes the identification of species as well as mating population, or biological species, whereas vegetative compatibility is used to identify the sub-populations or individuals within species. Mating compatibility of heterothallic microorganism could be the evidence of sexual fertility between two individuals in form of production of basidiospores, clamp connections and ascospore (Brasier, 1987).

Vegetative compatibility helps to understand genetic relation of the strains of fungi within a species (i.e., heterokaryon). It is simply on the ability of the mycelium of two fungal strains to anastome and is under the genetic control of multiple loci in the microorganism. It involves the confluent hyphal growth between two vegetative compatible strains. Vegetative compatibility is an important technique for the identification of genetic diversity within the fungal microorganism. Puhalla (1985) used the technique and classified different strains of *Fusarium*. He used "*nit*" mutant method for the vegetative compatibility experiment which involves the generation of *nit* mutants

organism. RT-PCR involves the use of SYBR Green dye to detect the differences in a specific genus of micro-organism.

Nucleic Acid Hybridization Based Detection

It is a recent molecular method and it offers an advantage over biochemical, i.e., serological detection methods. Nucleic acid hybridization based detection is based on high degree of specificity inherent in the pairing of nucleotide base sequences. This molecular detection tool is well stabilized, rapid and specific for the detection of the soil micro-organisms pathogenic to vegetable crops. It involves the labeling of nucleotide sequence and cloning of DNA strands of specific/desired pathogen/soil micro-organism. A DNA cloned from RNA strands is known as complimentary DNA (c DNA). With the help of transcription vectors, the hybrids of DNA: RNA is formed which is more stable than DNA:DNA hybrids and this may give clear assay for the detection of micro-organisms (Miller and Martin, 1988).

Development

Development of specific soil micro-organism help it cause a specific disease in a particular crop. Several factors like existence of host plant, presence of the specific micro-organism/pathogen and conducive environmental/soil conditions are pre-requisites for disease development. Environmental conditions are the most effective factors for development of particular soil micro-organism. Environmental factors like temperature, humidity, rainfall, wind, moisture, light, soil texture, pH, aeration and nutritional status of the soil of the particular location play a key role for the development of a particular soil micro-organism.

Most of the soil micro-organisms favour relatively high temperature, i.e., (78-90°F) for their development. Poorly drained soil and slight acidic conditions are best for the growth of *Fusarium* and *Verticillium* species are able to cause infection at low temperature, i.e., 70-80°F for disease development (Anon., 1997). Application of more nitrogen and less potassium fertilizers favours all soil-borne fungi to grow faster. *Ralstonia* species like warm and humid conditions with high temperature for their growth.

Another factor for the development of pathogenic soil micro-organisms is their mode of over-wintering in the soil. As it is known all the soil micro-organisms which are pathogenic to vegetable crops survive in soil for long time by means of different specialized structure such as spores, sclerotia, chlamydospores and conidia (Agrios, 2004).

Fusarium, the most important soil micro-organism, infectious to a large number of vegetable crops, survive in the soil in the form of three types of asexual spores *viz-*

microconidia, macrocinidia and chlamydospore. Microconidia is single celled or double celled, while macrocinidia are 3-5 celled structure. Chlamydospores are 1-2 celled, roundish, and thick walled structures.

Verticillium species produce one celled, colourless conidia that are short lived but microsclerotia produced by the species are able to survive in soil for more than 15 years. Mycelium of *Pythium* produces sporangia which stay in soil for a long time and are able to cause infection by releasing the zoospores. It also produces oospores which in favorable conditions produce zoospores and cause infection in vegetable crops.

Ralstonia, the bacterial pathogen survives in the soil, seed as well as in the insect vector in the form of oozes. When they come in contact with host plant they enter into the plant system through wounds, multiply in the vascular system (xylem) and cause blocking of the vessels.

MANAGEMENT

Disease management requires a thorough knowledge on three main components such as host, pathogen and environment. The disease development and progress depends on the interaction of all three factors. For a successful disease development the host must be susceptible interacting with a virulent pathogen in favorable environmental conditions. When disease management practices are applied, the value of the crop must exceed the cost of control practices. Hence possible approaches like use of resistant/tolerant varieties, cultural, chemical and biological methods must be integrated in wilt disease management. Solo application of any of the above management method is unable to solve disease problems completely, because each is dually endowed with benefits and drawbacks. Integrated disease management or IDM is required for a sustainable management of diseases. The aim of application of an IDM program is to get benefits in vegetable production at low cost and low risk. The process of application of appropriate control measures in IDM program is discussed as under:

A. Cultural Control Methods

An application of suitable cultural control strategy is helpful in integrated disease management (IDM) programme. By the application of suitable cultural practices the disease problem can be reduced and simultaneously increase vegetable production.

Selection of Planting Site

The vegetable production site should be slightly sloped to provide good water drainage through the soil. Excess soil moisture can damage vegetable roots and promote root diseases caused by soil microorganisms like damping off of vegetables caused by a complex of *Pythium-Rhizoctonia-Sclerotiums* pp.

Selection of Disease Free and Resistant/Tolerant Varieties

Selection of disease free seed and planting material is one of the most important aspects in IDM to improve vegetable production. Seeds should be obtained from recognized organization because the external appearance of the seeds might not be indicative about the seed health. Use of resistant variety is pivotal in controlling several vegetable diseases, e.g., seedling blight of tomatoes

Sanitation and Crop Rotation

Several soil microorganisms can survive in plant debris for several years and "sanitation "is a key tool in controlling these microorganisms. Sanitation involves removal or plowing-under of crop stubble and trash which helps to destroy the population of disease causing microorganisms.

Crop rotation is a method for managing several soil-borne diseases and it helps to prevent the buildup of disease-causing microorganisms in the soil. A period of at least a three-year of crop rotation may be useful for vegetable crops to minimize the population of soil-borne pathogens.

Proper Use of Irrigation and Fertilizers

It is reported that wet foliage is favourable for the development of most of the diseases. To reduce infections, irrigation must be done in late morning or mid-afternoon to allow the foliage to dry before evening. Maintaining uniform soil moisture can reduce problems such as blossom end rot of chillies and tomatoes. Application of fertilizers should depend upon the soil character (acidity/alkalinity) and nutrient status. So before application of fertilizers soil should be tested and fertilizer may be applied accordingly.

Removal of Infected Plant/Plant Parts

Removal of infected plant and plant parts carefully to inhibit the spread of pathogen inoculum all around the site is an important step of vegetable IDM.

B. Mechanical Control Methods

This method is an integration of mulching, and soil disinfestation which is effective in reducing soil-borne inoculum.

Mulching

Covering of ground with plastics or straw is called mulching. Plastic mulch is a widely used ground cover in vegetable production. It acts as a barrier between soil and plant surfaces and also helps in the reduction of the amount of disease inoculum splashed onto foliage, stems, and fruits during rainy periods or irrigation.

Soil Disinfestations

Soil disinfestation is an effective method for the removal of several destructive inoculum present in the soil debris, and the three approaches of soil disinfestation are

a) Steaming
b) Fumigation
c) Soil solarization.

Control of plant diseases through cultural practices involves the principles of avoidance, exclusion and eradication. Under cultural control method the use of resistance cultivar or hybrids offers the best possibility of control of *Fusarium* wilt. The temperature of solarized soil reached up to 52°C at depth of 10-15 cm (Abdul Wahid et al., 2001). Similarly at 15 and 25-cm depth, solarization caused an increase in soil temperature of 15.4 ± 2.7°C and 14.7 ± 1.84°C respectively. Soil solarization delayed the appearance of all the wilt diseases tested and reduced the incidence of *Verticillium* wilt of brinjal by 89 - 96% and melon *Fusarium* wilt by 59-94.6% (Tamietti & Valentino, 2000). In case of bacterial wilt diseases, practices like destruction of diseased plant, prevention of leaf cutting or animal browsing in infested plots, disinfection of pruning knives, selection of planting materials from disease-free field and crop rotation with non-host crops for two years were reported to be effective (Yirgou and Bradbury, 1974).

C. Biological Control Methods

Biological control methods involve the use of any bio-control agent/organism to regulate a pathogen population below its economic threshold level. So, biological control agent (BCA) can be defined as any organism that can suppress the pathogen. The utilization of BCAs is best advocated in soils which are rich in well decomposed organic matter, well moistened and protected from extreme weather conditions.

Biological control of any organism/pathogen resulted from different types of interactions between BCA and the pathogen/causal agent of any specific disease. The mechanisms by which biological control can be achieved are:

a) Direct antagonism
b) Mixed path antagonism and
c) Indirect antagonism

Biological control helps in the development of selected microorganisms which are beneficial to plant environment but antagonistic to pathogens. Numerous microorganisms such as non-pathogenic strains of *F. oxysporum*, *Bradyrhizobiumjaponicum*, *Pseudomonas* like *P. putida*, *P. fluorescens*, *Serratiamarcescens*, *Talaromycesflavus* and *Streptomyces* spp. were isolated and used to control *Fusarium*, *Rhizoctonia*, *Pythium* and *Verticillium* wilt of vegetables (Liu et al., 1995; Fravel et al., 1995; El-Abyad et al., 1993; Yamaguchi et al., 1992). The nonpathogenic *F. oxysporum* was effective as a biological control agent against wilt in solanaceous crops (Yamaguchi et al., 1992). Other organisms, including isolates of *Gliocladiumvirens*, *T. hamatum*, *P. fluorescens* and *Bacillus cepacia*, also significantly reduced *Fusarium* wilt (30 to 65% reduction), *Pythium*, *Sclerotinia*, *Sclerotium* and *Rhizoctonia* diseases. *T. flavus* has shown greater bio-control potential against *Verticillium* wilt diseases through production of the extra cellular enzyme glucose oxidase and the subsequent release of hydrogen peroxide (Larkin et al., 1998).The antagonistic *P. fluorescens* strain and the chitinolytic *S. Plymuthica* strain were also reported to be effective against *V. dahliae* (Kurzeet al., 2001). Six bacterial strains viz. *Corynebacterium sp.*, *Bacillus* spp., *Escherichia* sp., *Serratia* sp. and *Pseudomonas* sp. were found most effective against bacterial wilt (*R. solanacearum*) (Guo et al., 2001). *B. subtilis* was found effective against *Fusarium* wilt of tomato and other vegetables (Sarhanet et al., 2001). Several antagonistic bacteria like *Lysobactorenzymogenes* (OH 11), *P. fluorescens* (BOH 3), and *Bacillus* spp., were reported to control both fungal and bacterial wilt pathogens (Jiang et al., 2005). Similarly, *Penicilliumoxalicum* was also reported to reduce 22%-64% of wilt disease in tomato (Sabuquillo et al., 2006).

D. Chemical Control Methods

Uses of chemicals are one of the easiest and most suitable measures in the control of several vegetable pathogens in an IDM program. It acts as a direct protection between host and pathogen. Chemical control involves the use of several chemicals, i.e., synthetic chemicals and phytochemicals to manage the vegetable diseases.

Synthetic chemicals comprise of a broad range of chemicals with curative actions, like fungicides, bactericides and antibiotics which are used to prevent the infection of a wide range of pathogens. Chemicals may be applied by spraying, dusting, seed treatment and soil treatment. *In vitro* experiment with fungicides against *F. oxysporum* indicated that prochloraz and carbendazim were the most effective fungicides in controlling the pathogens (Song et al., 2004). In a field study, the tuber treatment with captan at 75 g a.i./kg in potato (cv. Kennebec) showed greater efficacy in controlling the diseases caused by *Fusarium spp., Rhizoctoniasolani, Streptomyces scabies* and *Verticillium spp.* (Platt and MacLean, 1997). Propineb was reported to be most effective against wilt pathogens of beans (Mukherjee & Tripathi, 2000). New fungicide combinations like fluopyram+tebuconazole, trifloxystrobin+tebuconazole, benalaxyl+mancozeb had been found effective against the major diseases of tomato, chillies and cucurbits (Saha et al., 2014; 2016; 2017) Several antibiotics were tested against *R. solanacearum* in which streptomycin and streptopenicillin recorded less bacterial infection (Singh et al., 2000).

CONCLUSION

Soil borne microorganisms, with their diverse nature are always a matter of intrigue to scientists. The detection, development and management of soil borne plant pathogens is of key importance in agricultural sciences as their groups have a major role to play in causing disease symptoms. The effective management strategies are yet to be deciphered and hence soil-borne pathogens are a continuous threat to sustainable agriculture. The deleterious effect of the soil borne microorganism must be negated with a view to obtain safe food and agricultural products.

REFERENCES

Abdul-Wahid, O. A., Moustafa, A. F. and Ibrahim, M. E. (2001). Integrated control of tomato Fusarium-wilt through implementation of soil solarization and filamentous fungi. *Zeitschrift fur Pflanzenkrankheiten und Pflanzenschutz [J. Plant disease and protection].*108 (4): 345-355.

Agrios, G. N. (2004). *Plant pathology.* Fifth edition. Elsevier Academic Press. Pp-922.

Anonymous, (1997). *Verticillium wilt diseases.* Department of crop science. University of Illinois at Urbana-Champaign. RPD. No. 1010, July 1997.

Broders, K. D., Wallhead, M. W., Austin, G. D., Lipps, P. E., Paul, P. A., Mullen, R. W. and Dorrance, A. E. (2009). Association of soil chemical and physical properties with *Pythium*species. Diversity, community, composition and disease incidence. *Phytopathology.* 99: 957-967.

Brasier, C. M. (1987). The dynamics of fungal speciation. Pages 231-260. In: *Evolutionary biology of the fungal mycelium.* A.D.M. Rayner, C.M. Brasier and D. Moore, eds. Cambridge University Press.

Correll, J. C. (1992). Genetic, biochemical and molecular techniques for the identification and detection of soil borne plant pathogenic fungi. Pages 7-16. In: *Methods for research on soil borne phytopathogenic fungi.* L.L. Singleton, J. D. Mihail and C.M. Rush, eds. APS Press, USA.

Douglas, S. M. (2008). *Verticillium wilt of vegetables and herbaceous ornamentals.* http:/www.ct.gov./case.

El-Abyad, M. S., El-Sayed, M. A., El-Shanshoury, A. R. and El-Sabhagh, S. M. (1993). Towards the biological control of fungal and bacterial diseases of tomato using antagonistic Streptomyces spp. *Plant and Soil.* 149: 1080-1086.

Fravel, D. R., Lewis, J. A. and Chittams, J. L. (1995). Alginate prill formulations of Talaromycesflavus with organic carriers for biocontrol of Verticillium dahlia. *Phytopathology.* 85: 165-168.

Galfre, G. and Milstein, C. (1981). Preparation of monoclonal antibodies: strategies and procedures. *Meth. Enzymol.* 73: 3-46.

Guo, J. H., Guo, Y. H., Zhang, L. X. and Fang, Z. D. (2001). Screening for biocontrol agents against Ralstonia solanacearum. *Chinese Journal of Biological Control.* 17 (3): 101-106.

Jiang, Y. H., Hu, B. S., and Liu, F. Q. (2005). Selection and identification of antagonistic bacteria against soil born plant pathogen. *Chinese Journal of Biological Control.* 21 (4): 260-264.

Jeffreys, A. J., Wilson, V. and Thein, S. L. (1885). Individual-specific fingerprints of human DNA. *Nature.* 316: 76-79.

Kohn, L. M., Petsche, D. M., Bailey, J. R., Novak, L. A. and Anderson, J. B. (1988). Restriction fragment length polymorphisms in nuclear and mitochondrial DNA of *Sclerotinia* species. *Phytopathology.* 78: 1047-1051.

Kistler, H. C., Bosland, P. W., Benny, U., Leong, S. and Williams, P. H. (1987). Relatedness of Fusarium oxysporum from crucifers measured by examination of mitochondrial and ribosomal DNA. *Phytopathology.* 77: 1289-1293.

Kurze, S., Sauerbrunn, N., Bahl, H. and Berg, G. (2001). Effects of antagonistic rhizobacteria on plant health, yield and the bacterial rhizosphere community of strawberry. *Bulletin-OILB/SROP.* 24 (3): 117-120.

Larkin, R. P., Fravel, D. R., Dugger, P. and Richter, D. (1998). Biological control of wilt pathogens with fungal antagonists. *1998 Proceedings Beltwide Cotton Conferences,* San Diego, California, USA, 5-9 January 1998. Vol. 1: 125-127.

Liu, L., Kloepper, J. W. and Tuzun, S. (1995). Inducton of systemic resistance in cucumber by plant growth promoting rhizobacteria: duration of proctection and effect of host resistance on proctection and root colonization. *Phytopathology.* 85: 1064-1068.

Micales, J. A., Bonde, M. R. and Peterson, G. L. (1986). The use of isoenzyme analysis in fungal taxonomy and genetics. *Mycotaxon.* 27: 405-449.

Miller, S. A. and Martin, R. R. (1988). Molecular diagnosis of plant disease. *Ann. Rev. Phytopathol.* 26: 409-432.

Mukherjee, S. and Tripathi, H. S. (2000). Biological and chemical control of wilt complex of Frenchbean. *Journal of Mycology and Plant-Pathology.* 30 (3): 380-385.

Platt, H. W., and MacLean, V. M. (1997). Efficacy of chemical control products for control of soil-borne potato diseases caused by soil-borne fungal pathogens in 1995. Tests of Agrochemicals and Cultivars. *Annals of Applied Biology.* 130 (18): 16-17.

Puhalla, J. E. (1985). Classification of strain of Fusarium oxysporum on the basis of vegetative compatibility. *Can. J. Bot.* 63: 179-183.

Sabuquillo, P., De, Cal. A. and Melgaregb, P. (2006). Biological control of tomato wilt by *Penicilliumoxalicum* formulation in different crop condition. *Biological control.* 37 (3): 256-265.

Saha, S., Shabeer, A., Purath, T., Jadhav, M. R., Loganathan, M., Banerjee, K. and Rai, A. B. (2014). Bioefficacy, residue dynamics and safety assessment of the combination fungicide trifloxystrobin25% + tebuconazole 50%-75 WG in managing early blight of tomato (*Lycopersiconesculentum* Mill.). *Journal of Environmental Science and Health, Part B: Pesticides, Food Contaminants, and Agricultural Wastes.* 49 (2): 134-141.

Saha, S., Jadhav, M. R., AhammedShabeer, T. P., Banerjee, K.,Bineet, K. Sharma, M. L. and Rai, A. B. (2016). Safety Assessment and Bioefficacy of Fluopyram 20% + Tebuconazole 20% - 40 SC in Chilli, *Capsicum annum* L. Against Anthracnose Disease. Proc. Natl. Acad. Sci., India, *Sect. B Biol.* 86 (2): 359-366.

Saha, S., Ashtekar, N.D., Rai, A.B. and Balaraman, A. K. (2017). Synergistic effect of Benalaxyl 8% and Mancozeb 65% WP in combating Downy mildew of Cucumber. *J. Mycopathol. Res.*55(3):227-230.

Sarhan, M. M., Ezzat, S. M., Tohamy, M. R. A., El-Essawy, A. A. and Mohamed, F. A. (2001). Biocontrol of *Fusarium* tomato wilt disease by *Bacillus subtilis. Egyptian Journal of Microbiology.* 36 (1): 103-118.

Singh, D. K., Sinha, S. K., Singh, V. N. and Singh, D. N. (2000). Control of bacterial wilt of ginger (Zingiberofficinale) with antibiotics. *Journal of Research,* Birsa Agricultural University. 12 (1):41-43.

Song, W. T., Zhou, L. G., Yang, C. Z., Cao, X. D., Zhang, L. Q. and Liu, X. (2004). Tomato Fusarium wilt and its chemical control strategies in a hydroponic system. *Crop-Protection.*23 (3): 243-247.

Tamietti, G. and Valentino, D. (2000). Effectiveness of soil solarization against soil-borne plant pathogens and weeds in Piedmont (northern Italy). *Acta-Horticulturae.* 532: 151-156.

Verma, L.R., Sharma, R. C. (1999). *Disease of Horticulture Crops. Indus publication comp.* New Delhi. Pp- 140-141.

Walker, J. C. (1957). *Plant Pathology.*McGraw Hill. Book Co. New York. Pp- 707.

White, T.J., Bruns, T., Lee, S. and Taylor, J. (1990). Amplification and direct sequencing of ribosomal RNA genes for phylogenetics. Pp. 315-322. In: *PCR Protocols*. M.A. Innis, D.H. Gelfand, J.J. Sninsky and T.J. White. Eds. Academic Press, NewYork.

Yamaguchi, K., Sano, T., Arita, M. and Takahashi, M. (1992). Biocontrol of Fusarium wilt of tomato and Verticillium wilt of eggplant by non-pathogenic Fusarium oxysporum MT0062. *Annals of the Phytopathological Society of Japan*. 58 (2): 188-194.

Yirgou, D. and Bradbury, J. F. (1974). A note on wilt of banana caused by the enset wilt organism Xanthomonas musacearum. *East-African-Agricultural-and-Forestry-Journal*. 40 (1): 111-114.

Selected Readings

Baayen, R. P. (2000). Diagnosis and detection of host- specific forms of *Fusarium oxysporum*. EPPO Bull. 30: 489-491.

Deacon, J. (2004). *Fungal Biology*. Blackwell Publishing Ltd. 4th edition. 293- 297.

Edel, V., Steinberg, C., Gautheron, N. and Alabouvette, C. (2000). Ribosomal DNA-targeted oligonucleotide probe and PCR assay specific for *Fusarium oxysporum*. *Myco. Res*. 104: 518-526.

Georgieva, O. (1992). Antagonistic characteristics of *Trichodermakoningii* towards Verticilliumdahliae on pepper. *Bulletin-OILB-SROP*. 15 (1): 18-20.

Gomez-Caicedo, L. E., Echeverry, N. E. and Gonzalez, S. R. (2001). Evaluation of cultural, chemical and biological control of vascular rot and wilt in plantain (Musa AAB Simmonds). *Infomusa*. 10 (1): 17-21.

Granada, G. A. and Sequeira, L. (1983). Survival of *Pseudomonas solanacearum* in soil, rhizosphere and plant roots. *Can. J. Microbiol*. 29: 433-440.

Lyons, N., Cruz, L. and Santos, M. S. (2001). Rapid field detection of Ralstoniasolanacearum in infected tomato and potato plants using the Staphylococcus aureus slide agglutination test. *Bulletin-OEPP*. 31 (1): 91-93.

Sood, A. K. (2005). Bacterial wilt of solanaceous vegetable crops and its management. In *challenging problems in horticulture and forest pathology*. (Edited by Sharma, R.C. and Sharma, J.N.) Indus Publication Company, New Delhi. 167- 182.

Steve, Bost. Plant Disease. *Tomato Wilt Problems*. http://www.utextension.utk.edu.

EDITORS CONTACT INFORMATION

N. C. Gautam, PhD
Professor in Agronomy and Vice-Chancellor
Mahatma Gandhi Chitrakoot Gramodaya University
Satna, MP, India
Email: gautamnc@yahoo.co.in

S. P. Tiwari, PhD
Associate Professor in Microbiology
Faculty of Science, VBS Purvanchal University
Jaunpur, UP, India
Email: sptiwarimicro@gmail.com

Rajesh Sharma, PhD
Senior Assistant Professor in Biotechnology
Faculty of Science, VBS Purvanchal University
Jaunpur, UP, India
Email: rajeshdbtpu@gmail.com

INDEX

A

B

C

D

E

F

G

H

I

O

P

Q

R

S

T

U

V

W

X

Y

Z

Recent Advances in Microbiology. Volume 1

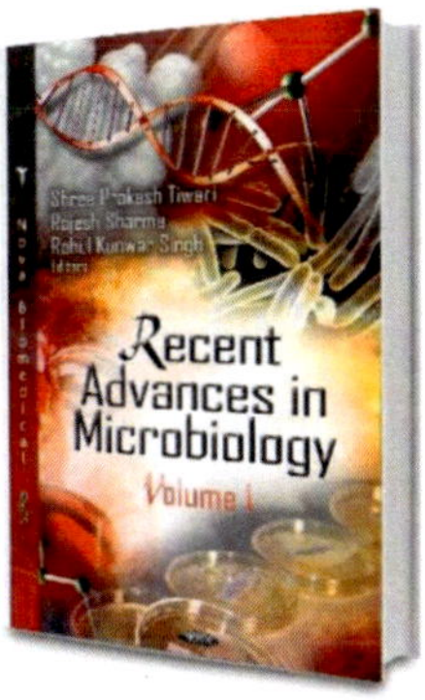

Editors: Shree Prakash Tiwari, Rajesh Sharma, (VBS Purvanchal University, Jaunpur, India) and Rahul Kunwar Singh (Banaras Hindu University, Varanasi, India)

Series: Recent Advances in Microbiology

Book Description: This book has been compiled with the help of leading Microbiologists from India and other countries with an objective of providing current developments in the field of Microbiology. This book will be useful for under and postgraduate students, academicians, researchers and laboratory personnel.

Hardcover ISBN: 978-1-61470-632-8
Retail Price: $295

Recent Advances in Microbiology. Volume 2

Editors: S.P. Tiwari, Rajesh Sharma and Rajeeva Gaur (Department of Microbiology, VBS Purvanchal University, Jaunpur, India, and others)

Series: Recent Advances in Microbiology

Book Description: To assist the readers in understanding the text, several illustrations and figures have been incorporated in the current volume. The topics covered in this volume are arranged in such a way as to maintain the thread of continuity with Volume One as well as to make it easy for those who are not acquainted with Volume One of the series.

Hardcover ISBN: 978-1-62808-628-7
Retail Price: $255

Microbiology and Biotechnology for a Sustainable Environment

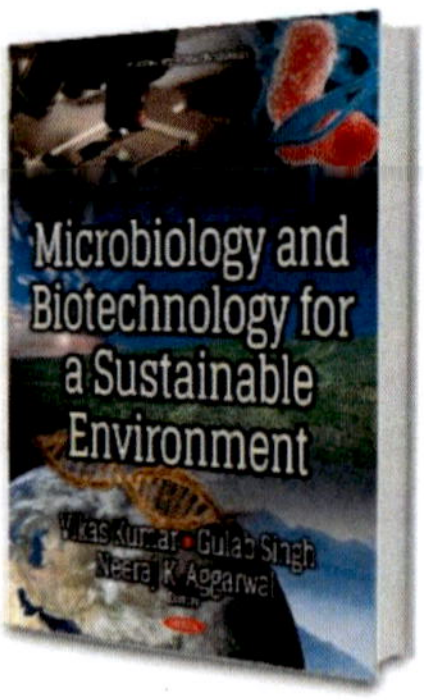

Editors: Vikas Kumar, Gulab Singh and Neeraj K. Aggarwal (Department of Biotechnology, M.M. University, Haryana, India, and others)

Series: Microbiology Research Advances

Book Description: The major goal of *Microbiology and Biotechnology for Sustainable Environment* is to emphasize all the disciplines from the subject of environmental microbiology and biotechnology. This book fills a long felt gap in the field of environmental sustainability and presents chapters authored by most active researchers in this area.

Hardcover ISBN: 978-1-53612-076-9
Retail Price: $230

Arbuscular Mycorrhizal Fungi: Effects, Analyses and Insights

Editors: Raul Bryan and Shirley Brooks

Series: Microbiology Research Advances

Book Description: In this book, the authors review their own studies on the impact of arbuscular mycorrhizal fungi (rhizospheric microorganisms) on plant host physiology and development. The increasing interest for the reduction of the input of chemical fertilizers and the use of combined microorganisms as biofertilizers, particularly applied in organic agriculture, is taken into account.

Softcover ISBN: 978-1-53612-981-6
Retail Price: $82